Glossary of United Kingdom
FISHING GEAR TERMS

Glossary of United Kingdom
FISHING GEAR TERMS

Compiled by

J P Bridger
J J Foster
A R Margetts
E S Strange

Fishing News Books Ltd
Farnham · Surrey · England

British Library CIP data

Glossary of United Kingdom fishing gear terms.
1. Fisheries—Equipment and supplies—
Dictionaries
I. Bridger, J. P.
639'.22'0280321   SH155

ISBN 0 85238 119 0

Published by
Fishing News Books Ltd
1 Long Garden Walk, Farnham
Surrey, England

Photoset by Paston Press, Norwich
Printed by Adlard and Son Ltd, Bartholomew Press,
Dorking, Surrey, England

Contents

Figures

Abbreviations used in text

BS   British Standard
Cyl   cylindrical
Eng   England
N Ire Northern Ireland
Ref   reference
Scot Scotland
Sph   spherical
Syn   synonym

Introduction

Throughout the United Kingdom the names and expressions used to describe or identify components of fishing gears often differ from region to region. Even in one locality or at one fishing port more than one name may be used for the same item. Sometimes two quite different items are known by the same name. For many of the terms no definition relating to fishing gear is to be found in an ordinary dictionary.

This glossary lists, in alphabetical order, over 1500 fishing gear terms known to be used currently by fishermen and their suppliers in England, Wales, Scotland and Northern Ireland. It was compiled from information collected over a period of 2–3 years by staff of the Ministry of Agriculture, Fisheries and Food's Fisheries Laboratory at Lowestoft and the Department of Agriculture and Fisheries for Scotland's Marine Laboratory at Aberdeen in interviews with fishermen, gear manufacturers and chandlers in their respective countries, and, for Northern Ireland, by written communication with the Aberdeen Laboratory.

The main purpose of this glossary is not only to present the names and terms as attributed variously to items of gear but also, by identifying preferred terms, to establish a reference standard name for each part of the most commonly used fishing gears. Whereas it is not intended that this should discourage the local use of synonyms, it is hoped that the preferred terms will be adopted for wider communication and thereby help to avoid misunderstanding between fishermen, manufacturers and suppliers, particularly in international trade and discussion.

Scope

The fishing gear covered by this glossary is limited to those items of fishing equipment that are in the water during the fishing process. Thus deck equipment such as winches is excluded, as are the ropes *etc* fixed to the vessel and used for shooting and hauling the gear, *eg* outhaulers, gilsons, trawl bag ropes. However, a few exceptions to this general rule do occur where the same name is given to two different items having very different functions, only one of which is in the fishing gear as defined above: in such cases both meanings of the term are given, *eg* 'messenger', the name for the rope used to pull the warps into the towing block when side-trawling, is also the common Scottish term for the rope along the length of a fleet of drift nets and used for hauling the nets.

Although some of the basic terms used for netting and in net making are covered by this glossary, the trade technical terminology for ropes, cordage and netting is excluded.

The glossary includes most of the terms in common use in the 1970s but excludes obsolete terms except for a very few which are identified as such.

Collection of information

Interviews were held with fishermen and chandlers in all the major fishing ports on the mainland of England, Wales and Scotland and with representatives of manufacturers of fishing gear. Island fishing centres and Northern Ireland were partially covered by postal survey only.

The conduct of interviews in England and in Scotland was broadly similar but differed in some respects. In Scotland most interviews with fishermen were conducted on board. They were asked to name the gear and components seen on the ship. For other gear, not on board, prepared diagrams and sketches were shown to the fishermen. These illustrations were also used to assist with interviews on shore and with manufacturers and suppliers. The names the interviewee used and brief information on his fishing or professional status and the type of fishing gear he uses or supplies were recorded at the time of each interview.

In England and Wales at first a similar method was used but because many of the fishermen were specialists with only relatively few sorts of gear they were often not the right persons to name parts of those gears they did not personally use. The prepared diagram method was therefore dropped so that any one interview was concerned with only those gears actually dealt with by the person being interviewed. With the gear in question, or parts of it, usually visible on the boat, on the quay or in the store, quick sketches were made of it, the various parts were named and notes were made as to interpretation. To avoid confusion, any visiting Scottish fishermen were deliberately not interviewed in English ports.

For the postal survey detailed illustrations, with each component numbered, were used. The recipient was asked to supply the names of those components familiar to him.

Compilation and presentation

All the terms used are listed alphabetically in the left-hand column of the glossary. 'Preferred' terms (see below) are printed in full capitals and other terms, synonyms, are in lower case. In the second column, for each synonym, is indicated the locality where the survey revealed that it is used, and in the third column is shown the type of gear to which each term (preferred or synonym) relates.

A fourth column contains either a description of the item that the term refers to and any synonyms (indicated by abbreviation 'Syn') or a cross reference to the preferred term. The fifth column, at the right hand side of each page, indicates the number of the illustration depicting that item of gear.

For some terms there is doubt about the correct spelling, while other terms, of which there is no doubt as to the correct spelling, are commonly and consistently mis-spelt. Where there are alternative spellings both are included in this glossary, but one has been preferred on the basis of common usage or etymology, *eg* SELVEDGE, selvage; BATING, baiting.

Provided that the interviewer was certain that at least one person normally used a term it has been included in the glossary, even though it may have been mis-used.

Under locality, the entry 'Common' denotes that the term, as defined in this glossary, was met with in at least three regions of both England and Wales combined, and Scotland. The entries 'Eng' for England and Wales or 'Scot' for Scotland mean that the term was used in at least three regions of that country but not in the other. Otherwise the locality where the synonym is used is identified more accurately by major port, by county, or by a rather wider area *eg* NE Eng.

The illustrations are labelled with only the preferred terms. They are intended to clarify the definitions and also to direct a reader to the preferred terms for a description of any gear. They are not intended to suggest design criteria. In general, the illustrations are grouped according to fishing method, but *Figures 1* and *2* deal with basic terms relating to netting and net-making.

There are a number of British Standards dealing with fishing nets, terminology for netting, cutting netting to shape, mounting and joining netting *etc*. Where applicable, terms used in these standards are shown in the glossary as preferred terms and the number of the standard, preceded by the abbreviation BS, appears at the end of the definition of the term.

Preferred Terms

Where, as is so often the case, there are a number of names for the same item, the compilers have selected one as being in some ways preferable to the others. It is this term that is taken as the standard to which synonyms are referred, is here called the preferred term and is printed in full capitals. It has been chosen with due regard to both its traditional use and the geographical extent of its use. Normally the preferred term is the one that was most commonly encountered on the survey, is readily understood by most fishermen (even if in conversation they more often used a local synonym), and is least likely to cause ambiguity. For parts of such gears as purse seine and ring net, which are far more frequently used in Scotland than in England and Wales, the preferred term is the one most commonly used in Scotland. Similarly, for primarily English gears, the preferred term is the one used in England. Where two names appeared to have equal claim to preference, the one judged to have the best technical sense or to be appropriate to a number of fishing methods has been chosen.

Errors and Omissions

Considerable attention has been paid to comprehensive coverage and accuracy in compiling this glossary. However, its very nature and the method of compiling it inevitably lead to some imperfections. For instance, for many gears and especially those favoured by inshore fishermen there are often, in many of the small fishing harbours, local names for the various items, and the nationwide coverage has not been able to take in all such harbours.

As fishermen and net riggers often move between ports it is not unusual to find a man at, say, Milford Haven who uses almost exclusively Lowestoft terms because he learned his trade at Lowestoft. Consequently a term may well be used in areas other than the area of which it is characteristic, and some terms may be more widely used than this glossary implies. So it is expected that some persons in the fishing industry will find the glossary, in some part, to be incomplete, or that they will disagree with some of its terminology. Suggestions for incorporation in any future revised edition will be welcomed by the publisher and by the authors.

Acknowledgements

The authors record their appreciation of the valuable help, time and advice given by fishermen, gear manufacturers and suppliers, and local fisheries officers during the survey collecting the terms included in this glossary.

Glossary of United Kingdom Fishing Gear Terms

Term	Locality	Gear Type	Definition	Fig
A-B DIRECTIONS		Net making	The directions parallel to a rectilinear sequence of mesh bars each from adjacent meshes. BS 4440:1974	1
Aft bracket	Grimsby	Trawl	LARGE BRACKET	
Aft triangle	W Scot	Trawl	LARGE BRACKET	
ALLOY BOBBIN		Danish seine Trawl	Light-weight groundrope bobbin usually drilled to allow flooding. See DRILLED BOBBIN	
Ammel	NE Scot	Trawl Danish seine	DAN LENO STICK	
ANCHOR		Danish seine	The main anchor holding the ship when anchor seining	26
ANCHOR BUOY		Danish seine	Large spherical inflated buoy, often covered with netting, supporting and marking the various ropes connected to the main anchor. *Syn* can buoy (NE Eng); mooring buoy (Grimsby)	26
ANCHOR CHAIN		Danish seine	Length of heavy chain between anchor and anchor wire	26
Anchor net	Moray Firth		SET GILL NET or SUNK GILL NET	
ANCHOR POT		Pots	Heavily weighted pot used at each end of fleet in place of end-stone	
ANCHOR ROPE		1 General	Main holding rope connected to anchor or anchor chain	
		2 Danish seine	Cable-laid rope acting as a spring between the anchor wire and anchor buoy. *Syn* anchor trot (Grimsby)	
		3 Dredge	Rope between anchor and the boat in haul-tow dredging which when wound in moves the boat and hence moves the dredge	

Term	Locality	Gear Type	Definition	Fig
Anchor shackle	Aberdeen	General	CABLE JOINING SHACKLE	
Anchor surface net	Moray Firth	Set gill net	SURFACE GILL NET	
Anchor trot	Grimsby	Danish seine	ANCHOR ROPE	
Angle	Granton Moray Firth	Trawl	BRACKET	
Angle iron	S Eng Scot	Trawl	BRACKET	
Angle iron chain	Aberdeen	Trawl	CHAIN BRACKET	
Angle iron shackle	Aberdeen	Trawl	BOW SHACKLE	
Angle shackle	Moray Firth	Trawl	BRACKET SHACKLE	
APRON		Pound net	Netting floor of bag net and stake net. *Syn* floor (NW Scot)	33
Arken	W Scot	Ring net	CORK	
Arm	1 W Scot 2 NW Eng	Beach seine Trawl	Combined wing and shoulder BUTTERFLY	
ARMOURING		Trammel net	Outer wall(s) of larger mesh netting. *Syn* outer net (W Scot); outer wall (S Eng); outwall, outwalling (S Eng) trancher (Cornwall); wall (Dorset); walling (S Eng); windows (SW Scot)	3
Axle	NW Scot	Trawl	DAN LENO SPINDLE	
BACK		Dredge	Section of netting forming top side of the bag. *Syn* cover (Scot); dredge net (Cornwall); net (W Scot, SW Scot); net bag (W Scot); netting top (N Wales); top netting (Moray Firth); twine net (Cornwall)	28
Back	1 Cornwall 2 NE Eng Moray Firth 3 Cornnwall 4 NE Eng 5 N Ire	Handline Longline Pots Drift net Trawl	CAST MAIN LINE BACK ROPE HEADLINE of a salmon drift net BATINGS	
Back bars	Hull	Trawl	CHANNEL PLATES	
Back board chain(s)	Granton	Trawl	CHAIN BRACKET	
Back channels	Grimsby	Trawl	CHANNEL PLATES	
Back cord	SW Scot	Trawl	HEADLINE of a beam trawl	
BACK LINE		Handline	Main line to end of which cast or paternoster is attached	32
Back net	S Eng	Trawl	Rear sections of belly and batings plus codend. See also lower end	

Term	Locality	Gear Type	Definition	Fig
BACK PLATE		Trawl	Central steel plate on the back of otter board	
Back of line	SE Scot	Longline	END ROPE	
Back of net	SW Scot	Trawl	Square and batings of a beam trawl made in one section	
Back board becket	Granton SE Scot	Trawl	BACKSTROP	
Backing	Cornwall SE Scot	Longline	MAINLINE	
Backing line	1 Devon 2 SW Eng	Longline Pots	MAINLINE BACKROPE	
BACKROPE		Pots	The main line connecting a number of pots. *Syn* back (Cornwall); backing line (SW Eng); bottom line (Wales); bush rope (NE Scot); groundrope (SE Scot, W Scot); leader (NW Scot, W Scot) main line (SE Scot, W Scot); messenger (Scot); stray line (S Eng); tow (W Scot). See also POT LINE	30
Backrope	1 Cornwall 2 W Scot	Drift net Ring net	HEADLINE HEADLINE	
Backstrap	Common	Trawl	BACKSTROP	
BACKSTROP(S)		Trawl	The wire(s), chain(s) or combination rope(s) between otter board and kelly's eye or bridle. *Syn* backboard becket (Granton, SE Scot); backstrap (common); board bridle (SE Scot); board leg (Aberdeen); board strop (Hull); door legs (Grimsby); door strop (W Scot); sling (S Eng)	16
BACKSTROP EQUALISER		Trawl	Single sheaved block and swivel used as a rolling coupling to a single wire in place of two backstrops. *Syn* backstrop roller (Aberdeen)	21
BACKSTROP LINK		Trawl	Triangular steel link with rounded corners on back of otter board to which the backstrop is shackled. *Syn* board link (Aberdeen); door sling ring (Aberdeen); shearboard link (Grimsby); VD link (Hull)	19
BACKSTROP NORMAN		Trawl	1 Special 'U' bolt to which backstrop is attached. *Syn* eye (Granton) 2 Loosely—any other arrangement for attachment of backstrop to otter board	19
BACKSTROP RING		Trawl	Steel ring on back of otter board for attachment of backstrop	19
Backstrop roller	Aberdeen	Trawl	BACKSTROP EQUALISER	
BAG		1 Danish seine	Main body of net between shoulders and codend	27
		2 Dredge	Part holding the catch	28
Bag	1 SE Scot SW Scot	Beach seine	BUNT	

(cont)

Term	Locality	Gear Type	Definition	Fig
	2 W Scot NW Scot	Pound net	FISH COURT	
	3 Scot	Purse seine	BUNT	
	4 Scot	Ring net, Trawl	BELLY AND BATINGS	
Bag becket	SE Scot NE Scot	Trawl	HALVING BECKET	
Bag becket leg	Hull	Trawl	HAULING LEG	
Bag becket rings	Hull	Trawl	HALVING BECKET RINGS	
BAG NET		Pound net	Moored floating salmon pound net	33
Bag string	E Scot	Trawl	BELLY LINE	
Bag strop	W Scot	Trawl	HALVING BECKET	
Bag wing	W Scot	Pound net	INNER SCALE	
BAIT		Line Pots	Any substance that by its appearance or smell is used to lure fish to take a hook or enter a pot or trap	29
BAIT BAND		Pots	BAIT STRING—usually of elastic rubber, eg from inner tube of car tyre. *Syn* bait rubber (Devon)	
Bait line	W Scot	Pots	BAIT STRING	
Bait rope	Moray Firth	Pots	BAIT STRING	
Bait rubber	Devon	Pots	BAIT BAND	
BAIT STICK		Pots	Wooden skewer used to secure bait in Cornish pot. *Syn* pitcher (Cornwall); preen (Cornwall); prime (Cornwall); skewer (Cornwall); skiver (Cornwall)	29
BAIT STRING		Pots	A string or cord doubled between cross plate and a bow or cross stick to hold bait in a pot. *Syn* bait line (W Scot); bait rope (Moray Firth)	29
Baiting	Common	Net making	BATING	
Baitings	Common	Trawl	BATINGS	
Balch	Fleetwood	Trawl	BOLSH	
Balchline	Fleetwood	Trawl	BOLSH	
Ball	1 NE Eng 2 Scot	Pots Danish seine Trawl	MARKER BUOY HEADLINE FLOAT	
Balsh	Fleetwood	Trawl	BOLSH	
Banana	Hull	Trawl	BUTTERFLY	
BANANA LINK		Trawl	Elongated steel link fitted to large bracket to retain bracket shackle. *Syn* checker (SE Scot)	21
BANDED BOBBIN		Trawl	Near spherical steel bobbin reinforced with a steel band round its vertical circumference	

Term	Locality	Gear Type	Definition	Fig
BANKING GEAR		Trawl	Short-bellied trawl with 'paravanes' mounted close to wing ends used on steep sand and gravel banks	
BAR		1 Net making 2 Dredge	One of four sides of a mesh. *Syn* leg (S Eng) Wood or steel bar across end of bag. *Syn* chain bag bar (W Scot); lifting bar (W Scot, SW Scot); wood bar (Cornwall)	1 28
BAR CUT		Net making	Cut in A-B direction	
Bar size		Net making	Length of mesh bar. A non-standard measure of mesh size. See MESH SIZE	
Bar triangle	Aberdeen	Trawl	BRACKET	
BARREL		Pound net	Wooden barrel used to float cable of a bag net	33
BARREL BUCKLE SWIVEL		Handline	Small swivel combined with quick release wire clip for attachment of trace, cast. *Syn* top swivel (Cornwall)	32
Barrel spacer	SE Scot Moray Firth	Trawl	RUBBER SPACER	
BARREL SWIVEL		Handline	Small swivel	32
BARREL SWIVEL WITH LEAD LINK		Handline	Small swivel with quick release clip for attachment of lead weight. *Syn* bottom swivel (Cornwall)	32
Basket	E Eng	Longline	STRING (one basket holds one string)	
Basketing sling	W Scot	Ring net	TOP BUNT	
Bass rope	Grimsby SE Scot	Danish seine Trawl	GRASS ROPE	
Bat	Common	Various	DAN LENO STICK	
Bat line	S Eng	Trammel net	BRIDLE	
Bath	Lowestoft	Longline	STRING, length of which fills a bath tub	
BATING		Net making	In hand braiding the result of knotting one mesh on to two adjacent meshes in the preceding row to reduce the width of the netting. *Syn* baiting (common); littling (Sussex); shrink (Dorset)	1
BATINGS		Trawl	Tapered section of top panel between square and extension piece. *Syn* back (N Ire); baitings (common); belly tops (S Wales); top bag (Scot); top belly (Moray Firth); top sheet (Moray Firth, SW Scot)	9 10 11 23
BEACH SEINE			An encircling net shot from a small boat and drawn ashore by ropes from each end of the net. *Syn* draught net (SW Scot); draw net (common); scringe net (W Scot, SW Scot); shore seine (S Eng); sweep net (SW Scot)	7

Term	Locality	Gear Type	Definition	Fig
Bead	Lowestoft	Trawl	Cylindrical wooden bobbin. See BOBBIN	
Bead bobbin	Lowestoft	Trawl	Cylindrical wooden bobbin. See BOBBIN	
BEAM		Trawl	Wood or steel spar which holds the net of a beam trawl open horizontally	23
Beam head	Devon	Trawl	TRAWL HEAD	
BEAM TRAWL			Trawl in which the net is held open by a beam, or spar, with a trawl head at each end	23
Beating	Scot	Longline	BEATING THREAD	
Beating net	Literature	Splash net	SPLASH NET	
BEATING THREAD		Longline	Hook whipping thread. *Syn* beating (Scot); hook thread (SW Scot); liner thread (SE Scot); mint (W Scot)	31
BECKET		Various	A loop of rope or wire for securing and/or enclosing some part of a fishing gear. *Syn* bicket (purse seine, Shetland). See also GROMMET	
Becket	Scot	Longline	DOUBLING	
Becket bobbin	Hull	Trawl	LANCASTER	
Becketed hook	NE Scot	Longline	Hook mounted to short monofilament cast with end loop for attachment to snood	
Beehive pot	S Wales	Pots	CORNISH POT	
Beehive rubber	Grimsby	Trawl	HALF SHAPE BOBBIN	
BELLY		Trawl	Section(s) of lower panel between lower wings and extension piece. *Syn* bottom bag (Scot); bottom belly (Hull); bottom sheet (W Scot); ground sheet (SE Eng); lower bag (SE Scot, W Scot); mid bag lower (SW Scot); sheet (W Scot); under-blade (S Eng)	9 10 23
BELLY AND BATINGS		Trawl	Belly and batings of a two panel trawl joined together along their side selvedges. *Syn* bag (Scot); pair of bellies (Hull, Fleetwood); top and bottom bag laced (Moray Firth)	9
Belly cord	SW Scot	Trawl	FISHING LINE of beam trawl	
BELLY LINE		Trawl	Strengthening rope from lower quarter to near-side lestridge, often extending to codend. *Syn* bag string (E Scot); quarter belly line (Aberdeen); rib line (Hull)	9
BELLY LINE CHAIN		Trawl	Short length of chain between fishing line at the quarter and groundrope. *Syn* bobbin quarter chain (Aberdeen); bosom chain (Granton)	
Belly tops	S Wales	Trawl	BATINGS	

Term	Locality	Gear Type	Definition	Fig
Bicket	Shetland	Purse seine	BECKET joining pursing ring to pursing ring bridle	
Big bracket	S Wales	Trawl	LARGE BRACKET	
Big heading	Cornwall	Drift net	GUARDING	
Big triangle	Granton	Trawl	LARGE BRACKET	
BIGHT/BIT	S Eng	Pound net	Roofless enclosure of a kettle net	
Bights	SE Scot	Ring net	STAPLES	
BLADE		Dredge	Inclined toothless steel bar forming lower margin of frame. *Syn* knife (Cornwall)	
BLADE RINGS		Dredge	Rings connecting chain belly to blade or tooth bar. *Syn* knife rings (Cornwall)	
BLINDER		Trawl	Small-mesh lining for codend	
B M V board	Hull	Trawl	OVAL OTTER BOARD	
Board	Common	Trawl	OTTER BOARD	
Board bridle	SE Scot	Trawl	BACKSTROP	
Board chain(s)	Scot	Trawl	CHAIN BRACKET	
Board leg	Aberdeen	Trawl	BACKSTROP	
Board link	Aberdeen	Trawl	BACKSTROP LINK	
Board strop	Hull	Trawl	BACKSTROP	
Board strop legs	Hull	Trawl	The two otter board attachment sections of a Y-shaped backstrop	
Board strop tail	Hull	Trawl	The single part of a Y-shaped backstrop	
Board swivel link	Granton	Trawl	SWIVEL TOWING CHAIN	
Board towing chain	Aberdeen	Trawl	SWIVEL TOWING CHAIN	
Boarder	Moray Firth	Drift net	GUARDING	
Boat-end buoy	Cornwall	Longline	Marker buoy or dan marking the last section of main line shot	
BOAT NET		Drift net	of a fleet of nets, the one nearest to the boat. *Syn* stem net (Lowestoft)	5
Bobber	S Eng	Trawl	HEADLINE FLOAT	
BOBBIN		Trawl	1 Roller, numbers of which are threaded on wire of specified length to form part of a groundrope (see GROUNDROPE). Bobbins may be made of wood, steel, composition rubber, and cylindrical, spherical, hemi-spherical or elliptical in shape. *Syn* bead, bead bobbin (cyl. wood, Lowestoft); bobbin wheel (cyl. wood, Granton, W Scot); cartwheel (cyl. rubber, Grimsby); disc bobbin (cyl. rubber, Moray Firth); full shape (sph. steel, Hull); rubber	15

(cont)

7

Term	Locality	Gear Type	Definition	Fig
			roller (cyl. rubber bobbin, SE Scot); rubber wheel (cyl. rubber bobbin, common); wheel bobbin (cyl. rubber bobbin, Aberdeen, Moray Firth); wooden bead (cyl. wooden bobbin, Lowestoft) 2 Component of dan leno assembly	
Bobbin chain	Aberdeen	Trawl	BOBBIN LINKS	
Bobbin lifting rope	W Scot	Trawl	QUARTER ROPE	
BOBBIN LINK(S)		Trawl	Link(s) of various sizes and shapes used to connect groundrope to fishing line. *Syn* bobbin chain (Aberdeen). See DOUBLE BOBBIN CHAIN	15
Bobbin quarter chain	Aberdeen	Trawl	BELLY LINE CHAIN	
Bobbin rope	1 W Scot 2 W Scot SW Scot	Trawl Trawl	BOBBIN WIRE QUARTER ROPE	
Bobbin shackle	Aberdeen	Trawl	V-SHACKLE (for dan leno spindle)	
Bobbin wheel	Granton W Scot	Trawl	Cylindrical rubber bobbin. See BOBBIN	
BOBBIN WIRE		Trawl	Groundrope wire on which bobbins are threaded. *Syn* bobbin rope (W Scot)	14
Bobbins and legs	W Scot	Trawl	GROUNDROPE	
Bobleno	Granton E Scot NE Eng	Trawl	DAN LENO ASSEMBLY	
Boboleno	SE Scot	Trawl	BUMPER BOBBIN	
Body	SE Scot	Drift net	LINT	
Bog(s)	Hull	Trawl	HEADLINE FLOAT	
Bog line	Hull	Trawl	FLOAT ROPE	
Bolch	Common	Trawl	BOLSH	
Bolch line	Common	Trawl	BOLSH	
Boll	SE Scot	Pots	BULL	
BOLSH		Trawl	Rope attached along edge of lower wings and bosom netting for securing in bights to fishing line. *Syn* balch, balch line, balsh (Fleetwood); bolch, bolch line (common); bolsh line (W Scot); bosh line (S Eng); mounting rope (SE Scot)	9
Bolsh line	W Scot	Trawl	BOLSH	
Bolsh rope	W Scot	Purse seine	STAPLING	
Bolters	Devon	Longline	LONGLINE	
BOLTROPE		Trawl	Load-bearing rope fixed to length of	17

Term	Locality	Gear Type	Definition	Fig
			lestridge. *Syn* selvage rope (N Wales); selvedge rope (Granton, N Wales); side line (Sussex); strongback (Hull, Grimsby)	
Boltrope leg	Granton		MIDDLE LEG	
Boltrope spreader	Granton	Trawl	MIDDLE LEG	
Boltrope spreading wire	Aberdeen	Trawl	MIDDLE LEG	
Bools	Scot	Ring net	STAPLING	
Boomerang	NW Eng	Trawl	BUTTERFLY	
Bosh line	S Eng	Trawl	BOLSH	
BOSOM		Trawl	1 Middle section of headline, or fishing line, or groundrope. *Syn* busom, busum (Hull); middle (W Scot). See also CROWN, BOSOM PIECE, SIDE BOSOM PIECE. 2 Netting adjacent to rope sections in 1 above. *Syn* bosom head (of belly, Moray Firth)	9
Bosom and legs	W Scot	Trawl	GROUNDROPE	
BOSOM BOBBINS		Trawl	The middle bobbin section of a groundrope	14
Bosom chain	Granton	Trawl	BELLY LINE CHAIN	
Bosom gag	SE Scot	Trawl	QUARTER ROPE	
Bosom gusset	Moray Firth	Trawl	LOWER WING GUSSET	
Bosom head	Moray Firth	Trawl	BOSOM of belly	
BOSOM MESHES		Trawl	First row of meshes of bosom—often of stronger or double twine	
BOSOM PIECE		1 Trawl	Narrow section of stronger netting across forward edge of belly adjacent to fishing line	12
		2 Danish seine	Small section of netting behind centre of groundrope	27
BOSOM SECTION		Trawl	of groundrope, the section bordering the bosom	
BOSOM TICKLER		Trawl	Tickler chain attached at each end across bosom section of groundrope	
Boss hook	Moray Firth	Purse seine	PURSING RING	
Bottle	Grimsby W Scot	Trawl Danish seine	HEADLINE FLOAT	
Bottom	NE Eng	Drift net	LEAD-CORED LINE	
BOTTOM		Pots	Base of pot	
Bottom bag	Scot	Trawl	BELLY	
Bottom belly	Hull	Trawl	BELLY	

Term	Locality	Gear Type	Definition	Fig
Bottom bridle	1 W Scot 2 Granton W Scot	Trawl Danish seine	BOTTOM LEG BOTTOM SWEEP	
Bottom cheam	Cornwall	Pots	CANE HOOP, the lowest	
Bottom doubling	W Scot	Ring net	SOLE ROPE	
Bottom extension	W Scot	Danish seine	BOTTOM SWEEP	
Bottom fish tail	Moray Firth	Trawl	LOWER TOE	
Bottom gathering	W Scot	Trawl	LOWER QUARTER MESHES	
Bottom gill net	SE Scot	Set gill net	SUNK GILL NET	
Bottom goring	W Scot	Trawl	LOWER QUARTER or LOWER WING GUSSET	
BOTTOM LEG		Trawl	Wire or chain connecting groundrope to dan leno or bridle or otter board. *Syn* bottom bridle (W Scot); bottom spreader (Grimsby, Scot); bottom spreading wire (Scot); bottom sweep (W Scot); footrope leg (Hull, Grimsby); footrope wire (S Eng); groundrope leg (E Eng); groundrope spreader (Granton); groundrope spreading wire (Granton); lower leg (Lowestoft); lower spreading wire (Aberdeen); tow chain (Lowestoft, Devon); toe leg (Hull, Grimsby, Fleetwood); toe wire (Fleetwood)	16 17
Bottom line	Wales	Pots	BACKROPE	
Bottom nets	SE Scot W Scot	Ring net	LOWER BUNT	
Bottom panel	Grimsby	Trawl	Lower wings and belly of four panel bottom trawl joined	
Bottom sheet	1 Moray Firth 2 W Scot 3 Granton	Trawl Trawl Ring net	LOWER WING BELLY DEEPENING	
Bottom sling	Granton	Ring net	MIDDLE BUNT	
Bottom spreader	1 Scot Grimsby 2 Moray Firth	Trawl Danish seine	BOTTOM LEG BOTTOM SWEEP	
Bottom spreading wire	1 Scot 2 Moray Firth W Scot	Trawl Danish seine	BOTTOM LEG BOTTOM SWEEP	
Bottom sweep	W Scot	Trawl	BOTTOM LEG	
BOTTOM SWEEP		Danish seine	Rope joining bottom of dan leno to fishing line (or sole rope). *Syn* bottom	26 27

Term	Locality	Gear Type	Definition	Fig
			bridle (Granton, W Scot); bottom extension (W Scot); bottom spreader (Moray Firth); bottom spreading wire (Moray Firth, W Scot)	
Bottom swivel	Cornwall	Handline	BARREL SWIVEL WITH LEAD LINK	
BOTTOM TRAWL			Trawl designed to work on sea bed (or lake floor etc). *Syn* demersal trawl (common)	8
Bottom wing	Cornwall	Trawl	LOWER WING	
Boudoir	S Eng	Pots	PARLOUR	
Boulter	Cornwall	Longline	LONGLINE	
Bouncer	W Scot	Trawl	BUMPER BOBBIN	
BOW		Pots	Half-hoop support for the netting of a creel. *Syn* cane (W Scot); crouke (NW Scot); hoop (S Eng)	29
Bow	SE Scot NE Eng	Drift net	BUOY	
BOW SHACKLE		General	Shackle with curved sides. *Syn* angle iron shackle (Aberdeen); harp shackle (NW Eng)	24
Bow strop	SE Scot	Drift net	BUOY ROPE	
Bowl	NE Eng	Drift net, Set gill net	BUOY	
Box	W Scot	Trawl	(of four panel bottom trawl) SIDE PANEL	
Box trawl	Fleetwood	Trawl	FOUR PANEL TRAWL	
BRACKET(S)		Trawl	One, or a pair of, triangular shaped steel frames hinged to the front face of otter board, to which the warp is attached. *Syn* angle (Granton, Moray Firth); angle iron (S Eng, Scot); bar triangle (Aberdeen); iron (SE Eng); triangle (common)	19 20 21
Bracket clamp	Grimsby	Trawl	CLASP	
Bracket norman	Hull Grimsby	Trawl	CLASP	
BRACKET PLATE		Trawl	Steel plate on back of otter board through which pass bolts of the clasp securing the bracket	
BRACKET SHACKLE		Trawl	Heavy bow shackle for connecting warp to otter board. *Syn* angle shackle (Moray Firth). See BOW SHACKLE	
BRAID		Pots	Spiral band of wickerwork securing the ribs of a Cornish pot. *Syn* gale (Cornwall); ringing (Cornwall)	29
Braiding rope	W Scot	Drift net	HEAD CORD or FOOT CORD	
BRAILER		Ring net Purse seine	Dip-net used to transfer catch from net to ship	

Term	Locality	Gear Type	Definition	Fig
Breastropes	Grimsby	Trawl	STORM LINES of a midwater trawl	
BRIDLE(S)		1 General	One or more connected branch ropes (wire or fibre) which share a common load	
		2 Pound net	Forked rope between head or cleek pole and eek rope or mooring lines of a bag net	33
		3 Dredge	Wires connecting tow bar to warp	28
		4 Ring net	Extension of sweep leading to top corner of wing	7
		5 Set net	Rope(s) connecting end of net to the groundline or dan line. *Syn* bat line (trammel net, S Eng); net stray (tangle net, Cornwall); sweep (gill net, trammel net, Dorset)	5
		6 Trawl	(a) Otter Trawl	
			The rope(s) usually of wire, between otter board or backstrop(s) and net or dan leno or legs. *Syn* cable (Aberdeen, Hull, Granton); ground cable (Aberdeen, NE Eng); groundwire (N Wales); leno (S Eng); slack back (Devon); sweep (Scot)	16 17 18
			(b) Beam Trawl	
			Wires or chains connecting warp and trawl. *Syn* goosefoot (Grimsby). See CHAIN TOWING BRIDLE	23
Bridle	1 Grimsby Granton	Danish seine	SWEEP	
	2 Scot	Danish seine	DAN LENO	
BRIDLE GEAR	Eng	Trawl	Trawl with bridle system between otter board and net. *Syn* VD gear, Vigneron-Dahl gear (Eng)	
BRIDLE SYSTEM		Trawl	The system of ropes (of wire or fibre but principally of wire) and hardware all of which together connect the otter board to one side of the trawl	8 16 17 18
Buff	Lowestoft SE Scot	Various	BUOY	
Buff strop	Lowestoft SE Scot	Various	BUOY ROPE	
BULL		Pots	of creels—longitudinal wooden runner of open slatted wood base. *Syn* boll (SE Scot); runner (SE Scot); side piece (NE Eng)	29
Bull(s)	W Scot	Ring net Trawl	STAPLING	
BULLDOG GRIP		General	U-bolt with specially shaped sleeve used to clamp together two wire ropes. *Syn* dog (S Eng)	25
Bullet	W Scot	Trawl	BUMPER BOBBIN	
Bulties	Cornwall	Longline	LONGLINE	
BUMPER BOBBIN		Trawl	Bobbin threaded on to the net end of a lower leg or bridle. *Syn* boboleno (SE Scot); bouncer (W Scot); bullet (W Scot);	16 17

Term	Locality	Gear Type	Definition	Fig
			stotter (W Scot, SW Scot); toe bobbin (Moray Firth); torpedo (W Scot); wing bobbin (W Scot)	
BUNT		1 Beach seine Purse seine Ring net	Section of stronger netting often loosely hung to form a bag or pocket, where the catch is aggregated when the net is hauled. *Syn* bag (Scot); basketing sling (W Scot); bunt piece (Moray Firth); hose (beach seine, Dorset); middle nets (ring net, W Scot); sling (Granton, SW Scot)	6
		2 Trawl	The section of lower wing which is over-hung by the square. *Syn* bunt piece (Moray Firth); lower bunt (Moray Firth); lower wing bunt (Hull); shoulder (Moray Firth, NE Scot, SW Scot)	9 10 12
BUNT BOBBINS		Trawl	Bobbins in that section of groundrope bordering the bunt. *Syn* wing bunt bobbins (Moray Firth)	14
BUNT BRIDLE		Purse seine	Handling rope fastened to the bunt end of headline and the dan. *Syn* sweep rope (Moray Firth)	7
Bunt cork	SW Eng	Beach seine	CENTRE FLOAT	
Bunt piece	Moray Firth	Trawl	BUNT	
BUNT SECTION		Trawl	of groundrope, the section bordering the bunt	14
BUNT TICKLER		Trawl	Tickler chain attached to groundrope at forward end of bunt section	
BUOY		Various	Large float used to mark or support part of the gear. *Syn* bow (SE Scot, NE Eng); bowl (drift net and set net, NE Eng); buff (gill net, drift net and purse seine, Lowestoft, SE Scot); pallet (W Scot, N Ire); pellet (gill net and drift net, W Scot)	5 7
Buoy line	SE Scot	Various	BUOY ROPE	
BUOY ROPE		Various	Rope connecting buoy to that part of the gear being supported or marked. *Syn* bow strop (drift net, SE Scot); buff strop (set net, drift net and purse seine, Lowestoft, SE Scot); buoy line (general); creel end (pots, SE Scot); creeve end (pots, SE Scot); dan line, dan rope (common); dan string (S Eng); head (pots, SE Scot, SW Scot); lanyard (Cornwall); messenger (pots, Granton); pennant (pots, W. Scot); strap (drift net, N Ire); strop (drift net, Lowestoft, SE Scot); targle (drift net, Cornwall)	
Bushrope	1 Scot 2 NE Scot	Drift net Pots	MESSENGER BACKROPE	
Busom/Busum	Hull	Trawl	BOSOM	

Term	Locality	Gear Type	Definition	Fig
BUTTERFLY		Trawl	Part of dan leno assembly. L-shaped steel plate shackled between dan leno spindle and legs. *Syn* arm (NW Eng); banana (Hull); boomerang (NW Eng); dan leno arm (Grimsby); dan leno bracket (SW Scot); dan leno spreader (Aberdeen); devil's elbow (Cornwall); spreader bar (Aberdeen)	16
BUTTON		Pots	Sliding knot, button or band around bait band or bait string to secure bait. *Syn* grimmet (NE Scot); nipper (NE Eng); sliding knot (W Scot); slip knot (Moray Firth); stopper (W Scot)	29
C-link	Grimsby	Various	FALSE LINK	
CABLE		Pound net	Main anchoring rope of the leader or head of a bag net. *Syn* wire (NW Scot)	33
Cable	Aberdeen Hull Granton	Trawl	BRIDLE	
CABLE JOINING SHACKLE		General	Strong D-shackle with countersunk pin. *Syn* anchor shackle (Aberdeen); cable shackle (common); chain shackle (Moray Firth). See SHACKLE	
Cable keep	Hull	Trawl	STOPPER	
Cable shackle	Common	Trawl	CABLE JOINING SHACKLE	
Cadge	Aberdeen	Longline	GRAPNEL	
CAMBERED OTTER BOARD		Trawl	Otter board that is curved in fore and aft direction	20
Can	Aberdeen Grimsby W Scot	Trawl	HEADLINE FLOAT	
Can buoy	NE Eng	Danish seine	ANCHOR BUOY	
Can line	Grimsby	Trawl	FLOAT ROPE	
CANE HOOP	Cornwall	Pots	One of several horizontal hoops secured by vertical wire ribs. *Syn* cheam (Cornwall)	
Cane	W Scot	Pots	BOW	
Canner	W Scot	Ring net	SWEEP	
CARBINE CLIP			Link with spring loaded gate. *Syn* dog clip (Granton)	
CARBINE HOOK WITH SCREW KEEP		General	Link with hinged gate and lock nut. *Syn* screw link (Aberdeen)	
Cartwheel	Grimsby	Trawl	CYLINDRICAL RUBBER BOBBIN	
CAST		Handline	Terminal yarn, or strand, to which hooks are attached by short droppers. *Syn* back	32

Term	Locality	Gear Type	Definition	Fig
			(Cornwall); gut (Cornwall)	
CAST NET			Circular net with small lead weights around its perimeter that is hand cast to cover fish in shallow water	
CENTRE FLOAT		Beach seine	The middle float on the headline. Usually larger than the rest and of distinctive colour. *Syn* bunt cork (SW Eng)	
Centre wire	Aberdeen	Trawl	MIDDLE LEG	
CHAFER		Trawl	Replaceable material (hides, old netting etc) fixed to underside of codend for protection against chafing on the sea bed. *Syn* dolly (Cornwall); false belly (common). See also THRUMS	
Chafer	NE Scot	Danish seine	GRASS ROPE	
CHAFING PLATE		Trawl	Steel plate covering front or back of bottom plank(s) of an otter board. *Syn* check plate (Aberdeen); face plate (common); keel plate (W Scot); sand plate (Hull, Grimsby); scrub plate (Hull); shoe plate (Aberdeen); shoeing plate (Aberdeen)	19
CHAIN BAG	Granton SW Scot W Scot	Dredge	Bag with underside made of linked steel rings	28
Chain bag bar	W Scot	Dredge	BAR	
CHAIN BELLY		Dredge	Lower side of bag, made of linked steel rings. *Syn* rings (Moray Firth); wire bellies (Cornwall). See also RINGLES and STRUTS	28
Chain belly	Devon	Trawl	STONE MAT	
CHAIN BRACKET		Trawl	Chain used in place of bracket on an otter board. *Syn* angle iron chain (Aberdeen); back board chains (Granton); chain triangle (Aberdeen); board chain (Scot); towing chains (Granton, Moray Firth)	20
Chain bridle	Devon	Trawl Dredge	CHAIN TOWING BRIDLE	
Chain connecting link	W Scot	General	FALSE LINK	
CHAIN GROUND-ROPE		Trawl	Groundrope of bare chain. *Syn* frame chain (beam trawl, Devon)	
CHAIN HEART		Trawl	Chain used instead of wire for ground-rope on to which are mounted rubber discs or rounded with rope *etc*	
Chain lancaster	Moray Firth	Trawl	LANCASTER, also SPACER	
Chain link	1 SE Scot 2 W Scot	Trawl Trawl	FALSE LINK TOGGLE	

Term	Locality	Gear Type	Definition	Fig
Chain mat	Devon	Trawl	STONE MAT	
Chain shackle	Moray Firth	General	CABLE JOINING SHACKLE	
CHAIN TOWING BRIDLE		Trawl Dredge	Chain bridle connecting warp to frame of dredge or beam trawl. *Syn* chain bridle (Devon)	
Chain triangle	Aberdeen	Trawl	CHAIN BRACKET	
Chain wrapping	S Eng	Trawl	WRAPPING CHAIN	
Change	W Scot	Ring net	The join between wing and shoulder	
CHANNEL PLATES		Trawl	U section steel bracing bars on back of otter board. *Syn* back bars (Hull); back channels (Grimsby)	19
Chaver	Moray Firth NE Scot W Scot	Trawl	LIFTING BAG	
Cheam	Cornwall	Pots	CANE HOOP	
Check plate	Aberdeen	Trawl	CHAFING PLATE	
Checker	1 SE Eng	Longline	Eye spliced at ends of each section of main line for attachment of next section and also a weight	
	2 SE Scot	Trawl	BANANA LINK	
Choker	Moray Firth	Trawl	HALVING BECKET	
CLASP		Trawl	Shaped steel plate two of which secure each bracket to the otter board. *Syn* bracket clamp (Grimsby); bracket norman (Hull, Grimsby); triangle clip (Grimsby)	19
CLEAN MESHES		Net making	Selvedge meshes in the T-direction. *Syn* mash (Cornwall); pick up (Grimsby, Hull, SW Eng); take ups (Grimsby)	1
CLEEK		Pound net	The first compartment of bag net or stake net that fish enter. *Syn* outer court (W Scot)	33
CLEEK POLE		Pound net	Vertical pole supporting inshore corner of a bag net. *Syn* gable (SE Scot)	33
CLIP LINK		General	Oval link cut on one side so as to permit another link, similarly cut, to be coupled. *Syn* cut link (Granton); quarter clip (common); seine link (common); split link (Scot)	22
Clog	Hull	Trawl	KELLY PAD	
Coble	Cornwall	Drift net	CORK BUOY	
Cod-lashing	Scot	Trawl	CODLINE	
CODEND		Trawl Danish seine	Terminal part of net where catch collects. *Syn* net end (Granton); poke (S Eng, SW Eng); Tail end (Granton). See Swag bag, Money box	9 10 11
Codend becket	Moray Firth	Trawl	HALVING BECKET	

Term	Locality	Gear Type	Definition	Fig
Codend clip	Grimsby	Trawl	CODEND WEDGE	
Codend double net	Hull	Trawl	LIFTING BAG	
Codend gag	SE Scot	Trawl	HAULING LEG	
Codend gun	SE Scot	Trawl	CODEND WEDGE	
Codend lashing	SE Scot	Trawl	CODLINE	
Codend rope	Moray Firth W Scot NW Scot	Trawl	CODLINE	
Codend sleeve	Devon	Trawl	LENGTHENER, usually one made of small mesh netting	
Codend tier	W Scot	Trawl	CODLINE	
Codend tow	Moray Firth	Trawl	CODLINE	
CODEND WEDGE		Trawl	Quick release wedging device for securing the codline. *Syn* codend clip (Grimsby); codend gun (NE Scot); Holland clip (Aberdeen); wage (Moray Firth); wedge (N Ire)	
CODLINE		Trawl	Draw-rope taken through last row of codend meshes to close that end. *Syn* cod-lashing (Scot); codend lashing (SE Scot); codend rope (Moray Firth, W Scot, NW Scot); codend tier (W Scot); codend tow (Moray Firth); lacing string (N Eng); poke lashing (S Eng); poke line (Devon)	9
CODLINE MESHES		Trawl	Last row of meshes in codend through which codline is reeved	9
COIL		Danish seine	One section of hauling ropes, usually 120 fathoms long. See also SEINE ROPE	
COIR		Trawl Danish seine	Soft-laid coconut fibre rope. Term often used to describe a groundrope made from coir rope	
COMBINATION ROPE		General	Rope made of mixed wire and fibre yarns	
Connecting link	Aberdeen Moray Firth	Trawl	FALSE LINK	
COPE BAR		Trawl	Steel bar protecting top edge of top plank of an otter board. *Syn* top frame (Grimsby)	19
Corb	Cornwall	Pots	STORE POT	
Cord	1 W Scot 2 Grimsby	Drift net Danish seine	FOOTCORD Heavy braided synthetic twine connecting netting of crown and shoulders to headline	
CORK		Various	Cylindrical float made of cork or plastic used on headline. *Syn* arken (W Scot)	

Term	Locality	Gear Type	Definition	Fig
Cork back	N Ire	Drift net	HEADLINE	
CORK BUOY		Drift net	Marker made of cylindrical corks threaded on rope. *Syn* cable (Cornwall)	
Cork rope	1 Common	Drift net Ring net Set net	HEADLINE—double or single	
	2 Moray Firth	Purse seine	FLOAT ROPE	
Corner	Moray Firth	Trawl	QUARTER or WING GUSSET	
CORNISH POT			Round based, beehive shaped pot made of withies or wire with top entry. *Syn* beehive pot (S Wales); inkwell pot (N Wales); lobster pot (Cornwall); withy pot (Cornwall)	29
Cotton	Moray Firth SW Scot	Longline	TIPPING	
Coupling hook	Grimsby	Trawl	G-HOOK	
COVER		Pots	Netting covering the framework of a creel	29
Cover	1 Scot 2 NW Scot 3 SW Scot 4 Aberdeen Moray Firth NE Scot	Dredge Pound net Trawl Trawl	BACK ROOF SQUARE 1 LIFTING BAG 2 SMALL MESH COVER	
Crave	SE Scot	Pots	CREEL	
Crazy mesh	Grimsby	Net making	DOUBLE FLY MESH	
CREASING		Net making	In hand braiding, the result of braiding two meshes on to one mesh in the previous row to increase width of netting. *Syn* make (Dorset); making (SE Eng); stolen mesh (Cornwall)	1
CREEL		Pots	Rectangular, or square, wooden based pot with semi-circular hooped framework covered with netting and entrance(s) at side(s) or end(s). *Syn* crib net (NE Eng); crave (SE Scot); creeve (SE Scot); crib (SE Scot); Scottish creel (Eng)	29
Creel end	SE Scot	Pots	BUOY ROPE	
Creel in	SE Scot	Pots	EYE	
Creeve	SE Scot	Pots	CREEL	
Creeve end	SE Scot	Pots	BUOY ROPE	
Crib	SE Scot	Pots	CREEL	
Crib net	NE Eng	Pots	CREEL or POT	
Cringle	Aberdeen Granton	Trawl	TOGGLE	

Term	Locality	Gear Type	Definition	Fig
Criny	E Eng	Pots	FUNNEL, in iron whelk pot, made of slack netting	
Cross piece	E Eng	Pots	CROSS PLATE	
CROSS PLATE	SE Scot	Pots	of creel, one of a number of wooden battens spanning the two bulls. *Syn* cross piece (E Eng); latt (NE Eng); spake (SE Scot)	29
CROSS STICKS		Pots	Longitudinal bars (wood or metal) fixed to the bows of a creel. *Syn* headsticks (NE Scot); overliers (SW Scot); rails (NE Eng); scawbs (NE Eng); side sticks (SE Scot, SW Scot); stretchers (SW Scot); strutting (S Eng); top sticks (SE Scot)	29
CROSSOVER		Trawl	Strengthening rope for double codend, crossing from inside lestridge of one codend to outside lestridge of the other	
Crouke	NW Scot	Pots	BOW	
Crouping	W Scot	Trawl	QUARTER MESHES	
CROWN		1 Danish siene	A small section of stronger netting behind the centre of the headline. *Syn* gusset (N Ire); mat (NE Scot); mouth piece (Hull)	27
		2 Trawl	Stronger section of netting across forward edge of square, and adjacent to headline. *Syn* mat (NE Scot)	12
Cut knot	Dorset	Net making	POINT	
Cut link	Granton	Trawl	CLIP LINK	
Cut mesh	E Eng	Net making	POINT	
CUT-OFF		Trawl	One of three handling ropes connected in series as first, middle and last cut-off to close, respectively, start of lengthener, centre of lengthener and codend of a midwater trawl. See also HALVING BECKET	
Cut off	Devon	Trawl	HALVING BECKET	
CUT SPLICE	Eng	General	The ends of two ropes overlapped and spliced into each other to form an eye	
CUTTING RATE		Net making	The sequential cutting of meshes to reduce netting width at a given rate	
CYLINDRICAL RUBBER BOBBIN		Trawl	Cylindrical bobbin of rubber or rubber compound. *Syn* cartwheel (Grimsby); disc bobbin (Moray Firth); rubber roller (SE Scot); rubber wheel (common); Teal bobbin (common); wheel bobbin (Aberdeen)	15
D-SHACKLE			Shackle with straight parallel sides (see SHACKLE)	24
Daffin	NE Scot	Drift net	NORSEL	
Dahn	Common	Various	DAN	

Term	Locality	Gear Type	Definition	Fig
DAN		Various	Large float (slabs of cork, or inflated rubber or plastic) with ballasted centre pole surmounted by a flag or radar reflector and used as a principal surface marker. *Syn* dahn (common); dhan (common); flag (Danish seine, Scotland); flag end (pots, SE Eng); flaggy bow (long-line, pots, NE Eng); sticky bow (NE Eng); winkie (with flashing light, Scot)	5 7 26 30 31
DAN ANCHOR		Various	Anchor used to hold a dan at the extremity of set fishing gear. *Syn* grape (SW Eng)	5 30 31
Dan buoy	Wales	Pots	MARKER BUOY	
DAN LENO		Various	A device between a single bridle and two or more legs or other ropes to the net, aiding their separation. *Syn* bridle (Danish seine, Scot); polie (Danish seine, Moray Firth)	7 16 26 27
Dan leno arm	Grimsby	Trawl	BUTTERFLY	
DAN LENO ASSEMBLY	Grimsby	Trawl	Usually refers to the more complex form of dan leno used for deep sea trawling and consisting of spindle-mounted steel bobbin and butterfly. *Syn* bobleno (Granton, NE Eng, E Scot); dan leno rig (Hull)	16
Dan leno bracket	SW Scot	Trawl	BUTTERFLY	
Dan leno bridle	Grimsby	Beach seine Danish seine	DAN LENO STROP	
DAN LENO HOOP		Danish seine	Dan leno in the form of a hoop made from bent wood with short rigging ropes wired to the outer circumference. *Syn* dan leno ring (SW Scot); geer (Granton); hoop (Scot); hoop bridle (Grimsby); round dan leno (SE Scot); yoke hoop (Moray Firth)	27
Dan leno rig	Hull	Trawl	DAN LENO ASSEMBLY	
Dan leno ring	SW Scot	Danish seine	DAN LENO HOOP	
DAN LENO SCUTTLE		Trawl	An open hemispherical bobbin, some-times used in a dan leno assembly. *Syn* half bobbin (Hull); scuttle (Grimsby)	16
DAN LENO SHACKLE		Trawl	V-SHACKLE—used at each end of dan leno spindle	16
DAN LENO SPINDLE		Trawl	Steel spindle through dan leno bobbin or scuttle. *Syn* axle (NW Scot); spindle (common)	16
Dan leno spreader	Aberdeen	Trawl	BUTTERFLY	
DAN LENO STICK		Trawl Beach seine Danish seine	Dan leno in form of ballasted wood pole with short rigging ropes attached. *Syn* ammel (NE Scot); bat (common); pole	7 27

Term	Locality	Gear Type	Definition	Fig
			(Scot); spreader (W Scot, Cornwall); stick (common); stick dan leno (common); yoke stick, yoke (Moray Firth)	
DAN LENO STROP			The leading rigging rope of dan leno hoop or stick. *Syn* dan leno bridle (Grimsby); stick bridle (Grimsby)	7 27
Dan leno tail-piece	Aberdeen	Trawl	DAN LENO TRIANGLE	
DAN LENO TICKLER		Trawl	Tickler chain running from one dan leno to the other	
DAN LENO TRIANGLE		1 Trawl	Dan leno in form of a triangular steel plate. *Syn* dan leno tail-piece (Aberdeen); French dan leno (Grimsby); iron bridle (NE Eng); three bridle butterfly (Aberdeen); triangle (Moray Firth); triangle butterfly (Aberdeen)	17
		2 Danish seine	Dan leno in the form of an open triangular steel frame	27
DAN LENO WASHER		Trawl	Large washer used on spindle of dan leno, one each side of bobbin or scuttle. *Syn* plate (NE Scot); protective plate (Aberdeen); spindle ring (Aberdeen); washer (common)	16
DAN LINE		Various	Light rope connecting dan to anchor or weight. *Syn* dan string (S Eng); dan tow (E Eng); end (longline, pots, Moray Firth, SE Scot); end rope (pots, SW Scot, W Scot); end tow (longline and pots, W Scot, SW Scot); leader (longline, SE Scot); line end (Moray Firth); pennant (longline, W Scot)	5 30 31
Dan line	1 Scot	General	Sometimes describes size of line required for general purposes	
	2 Wales	Pots	BUOY ROPE	
Dan rope	Moray Firth SW Scot	Pots	BUOY ROPE	
Dan string	S Eng	Longline Pots	DAN LINE or BUOY ROPE	
Dan tow	E Eng	Longline	DAN LINE or BUOY ROPE	
DANDY		Handline	Comprised of up to four light-weight brass or stainless steel rods, centrally attached and spaced at right angles one above another on the weighted end of a handline from which baited hooks are suspended. *Syn* darra (Moray Firth); dodler (SW Scot); herring handline (NE Scot); jig line (SE Scot); jigger (SE Scot); sprool (W Scot); sprule (W Scot)	32
Dandy shackle	Scot	Trawl	D-SHACKLE WITH SQUARE HEADED SCREW PIN	

Term	Locality	Gear Type	Definition	Fig
Dangles	Lowestoft	Trawl	A series of steel rings surrounding groundrope with a form of tickler chain attached (obsolete)	
Dangling link	Grimsby	Trawl	DOUBLE BOBBIN CHAIN/LINKS	
DANISH SEINE			A seine incorporating a funnel-shaped net (with wings and codend) and very long ropes set out on the sea bed and hauled to a vessel in the open sea	26 27
Daphnes	NW Scot	Pots	POT STROP	
Darra	Moray Firth	Handline	DANDY	
Darrow	W Scot	Handline	SET OF FEATHERS	
Dead net	Sussex	Trawl	SQUARE	
Deep string	W Scot	Ring net	GABLE	
DEEPENING		Ring net Purse seine	Netting below shoulders and bag to deepen net. *Syn* bottom sheet (Granton); deeping (Granton); sheet (Granton)	6
Deeping	Granton	Ring net	DEEPENING	
Deepling	W Scot	Pound net	DOUBLING	
Demersal trawl	Common		BOTTOM TRAWL	
DEPRESSOR		Trawl	Shearing device (plane) fitted to lower rigging to generate a downwards pull on the lower front edge of a midwater trawl	
Devil's elbow	Cornwall	Trawl	BUTTERFLY	
Dhan	Common	Various	DAN	
Diamond	Hull	Net making	POINT	
Disc bobbin	Moray Firth	Trawl	CYLINDRICAL RUBBER BOBBIN	
Divided link	Moray Firth	Trawl	FALSE LINK	
DIVISIONAL BAR		Trawl	Longitudinal steel bar inserted between the planks of an otterboard with its ends bolted to the frame. *Syn* sandwich plate (W Scot); strengthener (Grimsby); through bar (Fleetwood); tie bar (Hull, Grimsby); trawl board bar (Aberdeen)	19
Dodler	SW Scot	Handline	DANDY	
Dog	S Eng	General	BULLDOG GRIP	
Dog clip	Granton	General	CARBINE CLIP	
Dog line	SE Scot	Trawl	LAZY DECKIE	
Dog tooth	Hull	Net making	DOUBLE FLY MESH	
Dog tow	Moray Firth	Trawl	LAZY DECKIE	
Dogrope	NE Eng Scot	Trawl	LAZY DECKIE	

Term	Locality	Gear Type	Definition	Fig
Dolly	Cornwall	Trawl	CHAFER	
DOODLER		Handline	Set of feathers with a ripper attached to lower end	32
Dooker	NE Eng	Pots	MARKER BUOY, cheaply improvised	
DOOR		1 Pots 2 Pound net	An opening for the removal of the catch See LARGE DOOR, SMALL DOOR	29
Door	Common	Trawl	OTTER BOARD	
Door chain	1 Aberdeen	Trawl	SWIVEL TOWING CHAIN; SWIVEL TOWING CHAIN WITH RECESSED LINK	
	2 Lowestoft	Trawl	OTTER BOARD TICKLER	
Door chain and swivel	Aberdeen Moray Firth	Trawl	SWIVEL TOWING CHAIN	
Door legs	Grimsby	Trawl	BACKSTROPS	
Door ring	Aberdeen	Trawl	LIFTING RING	
Door sling ring	Aberdeen	Trawl	BACKSTROP LINK	
Door strop	W Scot	Trawl	BACKSTROP	
Door swivel chain	Aberdeen	Trawl	SWIVEL TOWING CHAIN	
Door tickler	Lowestoft	Trawl	OTTER BOARD TICKLER	
Door to door tickler	Lowestoft	Trawl	OTTER BOARD TICKLER	
DORMANT MESHES		Gill net	Unsupported selvedge meshes between norsel attachments. *Syn* lazy meshes (Lowestoft); sleepy meshes (literature)	
Double bag becket	Hull	Trawl	HALVING BECKET	
Double bag becket extension	Hull	Trawl	LAZY DECKIE	
Double beam gear	Devon	Trawl	DOUBLE BEAM TRAWL	
DOUBLE BEAM TRAWL			Two beam trawls towed simultaneously. *Syn* double beam gear (Devon); Dutch double beam (E Eng); twin beam gear (Devon)	
DOUBLE BOBBIN CHAIN/LINKS		Trawl	Pair of large steel links, used to connect fishing line to groundrope. *Syn* dangling links (Grimsby); double dangle (Grimsby, Lowestoft)	15
Double bridle trawl	Eng	Trawl	TWIN BRIDLE TRAWL	
DOUBLE CODEND(S)		Trawl	Two codends in the form of a Siamese pair, joined together at the leading ends only. *Syn* split codend (Grimsby); trouser codend (Grimsby)	

Term	Locality	Gear Type	Definition	Fig
Double corner meshes	Hull	Trawl	QUARTER MESHES, of double twine	
Double dangle	Grimsby, Lowestoft	Trawl	DOUBLE BOBBIN CHAIN/LINKS	
DOUBLE FLY MESH		Trawl Danish seine	Fly mesh made of double twine. *Syn* crazy mesh (Grimsby); dog tooth (Hull)	
DOUBLE HEADLINE		Drift net Ring net	Headline made of two ropes of opposite lay. *Syn* set of headropes (Cornwall); top doubling (W Scot)	4
Double link	W Scot	Trawl	TOGGLE	
DOUBLE MESH		Net making	Mesh made of double twine	
Double mesh	SE Scot	Ring net	GUARDING	
DOUBLE SELVEDGE		Net making	Selvedge with edge meshes knotted into every row. *Syn* straight selvedge (literature)	
DOUBLING		1 Longline	Thin twine, doubled, connecting hook to snood. *Syn* becket (Scot); loop (W Scot)	31
		2 Pound net	Second net chamber leading to the fish court of a bag net or stake net. *Syn* deepling (W Scot); middle court (W Scot); middle pocket (Moray Firth); middle trap (SE Scot); outer bag (W Scot)	33
Doubling becket	Hull	Trawl	HALVING BECKET	
DOVETAIL	Aberdeen	Trawl	Crown and top wing gussets made in one piece	12
DRAG NET			Any fishing net which is operated by dragging it horizontally through the water	
DRAGLINK CONNECTOR			Articulated connecting link with central pin for chain and eye-spliced ropes	25
Draught net	SW Scot		BEACH SEINE	
Draw net	Common		BEACH SEINE	
DREDGE			General term for towed devices with rigid framed mouth for catching various species of bivalve shellfish	28
Dredge net	Cornwall	Dredge	BACK	
Dredge pole	SW Scot	Dredge	TOW BAR	
Dredge rope	Cornwall	Dredge	Towing rope between boat and oyster dredge	
DRIFT NET			Gill net that is free to stream and drift with wind and tide. *Syn* driving net (salmon, NE Eng)	4 5
DRILLED BOBBIN		Trawl	Hollow bobbin, usually spherical with a number of holes drilled through wall to allow free flooding	

Term	Locality	Gear Type	Definition	Fig
Driving net	NE Eng	Gill net	DRIFT NET (for salmon)	
Drop	NW Scot	Trawl	TOGGLE	
Drop chain	W Scot	Trawl	TOGGLE	
DROPPER		Handline	Short branch line connecting a hook to the cast	32
Dropper	NW Scot	Pots	POT STROP	
Drops	Moray Firth	Trawl	WING GUSSET	
Dryer	N Ire	Ring net	TOP BUNT	
Duker	NW Scot	Pots	MARKER BUOY, cheaply improvised	
Dumb string	1 Granton 2 SE Scot	Longline Pots	END ROPE END ROPE	
DUMMY	Hull Grimsby	Trawl	Iron spacer between bobbins with no provision for attachment to the fishing line	
Dummy	SE Scot	Pots	END ROPE	
Dutch double beam	E Eng	Trawl	DOUBLE BEAM TRAWL	
DUTCH POT		Pots	Upright cylindrical pot of steel framework covered with wire netting, entry at top end	29
EEK ROPE		Pound net	Rope section of a bag net mooring line for adjusting its length	33
Eel trap	Granton Moray Firth		FYKE NET	
ENCIRCLING NET			General term for a net used to surround a shoal of fish *eg* ring net, purse seine	6 7
End	1 Moray Firth SE Scot 2 SE Scot 3 W Scot	Longline Pots 1 Pound net 2 Ring net	DAN LINE DAN LINE HEAD SWEEP or WING	
END CHAFER		Trawl	Optional steel bar protecting leading and trailing ends of the otter board	
End cord	Cornwall	Drift net	GABLE	
End cording	Cornwall	Tangle net	GABLE—especially of ray nets	
End line	1 N Ire 2 NE Eng	1 Danish seine 2 Beach seine Trawl	WING LINE GABLE WING LINE	
End lining	Cornwall	Drift net	GABLE	
END PLATE		Trawl	Steel cladding welded to frame of an otter board in place of stiffeners to give added strength to leading or trailing ends	
END ROPE		Longline Pots	Line connecting end of first or last section of backrope or string to the dan line. *Syn* *(cont)*	30 31

Term	Locality	Gear Type	Definition	Fig
			back of line (SE Scot); dumb string (long-line, Granton; pots, SE Scot); dummy (pots, SE Scot); end tow (longline, NE Eng); lud tow (NE Eng); spreadline (Devon)	
End rope	SW Scot W Scot	Pots	DAN LINE	
END STONE		Pots	Stone, concrete or iron weight used to anchor each end of the string	30
End strap	Aberdeen	Trawl	STIFFENER	
End tow	1 NE Eng 2 W Scot SW Scot	Longline Longline Pots	END ROPE DAN LINE	
ENTANGLING NET			Loosely hung vertical net that catches fish by entangling rather than enmeshing, *eg* tangle net, ray net	
ENTRANCE		Pound net	Opening through which fish enter into the first chamber of a bag net or stake net. *Syn* first door (W Scot, SW Scot)	33
Entrance	W Scot	Pots	FUNNEL	
Extension	W Scot	Danish seine	SWEEP	
EXTENSION PIECE		Trawl	Tapered section of netting between belly and batings, and codend. *Syn* pipe (W Scot); swallow piece (beam trawl, SW Scot); tail (Moray Firth, W Scot); tail piece (Moray Firth); taper (Shetland); Y-piece (Granton)	9 10 11 23
EYE		Pots	Opening at the inner end of the funnel. *Syn* creelin (SE Scot). See also FUNNEL	29
Eye	1 Granton 2 Common 3 Common	Trawl General Pots	BACKSTROP NORMAN EYE SPLICE EYE and/or FUNNEL	
Eye bolt	Common	General	RING BOLT	
EYE LINE		Pots	Cord(s) that hold the funnel in position	29
EYE SPLICE		General	The end of a rope turned back and spliced into standing part to form a loop or eye. *Syn* checker (longline, SE Eng); eye (common)	
Eye triangle	Aberdeen	Trawl	LARGE BRACKET	
Face plate	Common	Trawl	CHAFING PLATE	
False bellies	Common	Trawl	CHAFER	
FALSE HEADLINE		Trawl	System of wires attached to either the otter boards or the dan lenos for the rigging of kites	13
FALSE LINK		General	A joining link that comprises two opposite C-shaped parts, one with riveting studs that locate in holes in the other part. *Syn*	25

Term	Locality	Gear Type	Definition	Fig
			C-link (Grimsby); chain connecting link (W Scot); chain link (SE Scot); connecting link (Aberdeen, Moray Firth); divided link (Moray Firth); split connecting link (Aberdeen); split link (Scot)	
Fancy line	SE Eng	Trawl	LAZY DECKIE	
FASTENING		Pots	Cord through edge meshes of cover, to facilitate securing cover to base. *Syn* lacing (NE Eng)	29
Feathered serpent	SE Scot	Handline	SET OF FEATHERS	
Feathers	Cornwall	Handline	SET OF FEATHERS	
Figure of eight link	Grimsby	Trawl	STOPPER	
FILLERS		Pots	of Cornish wire pot, vertical wires inserted between ribs and reaching only half way up the pot. *Syn* forcers (Cornwall)	
First door	W Scot SW Scot	Pound net	ENTRANCE	
FIRST POUND		Kettle net	Enclosure nearest to the shore	
FIRST ROUND		Net making	of half meshes. A series of loops fixed to a line or bar by clove hitches at the commencement of a piece of hand braided netting. *Syn* overcome (S Eng); overing (S Eng)	
FISH COURT		Pound net	The last net chamber of a bag net or stake net. *Syn* bag (W Scot, NW Scot); fish head (W Scot)	33
Fish door	W Scot	Pound net	LARGE DOOR or SMALL DOOR	
Fish head	W Scot	Pound net	FISH COURT	
Fish hook	Common	Handline	HOOK	
Fish line	N Ire	Trawl	FISHING LINE	
Fish rope	SW Scot SE Scot	Trawl	LAZY DECKIE	
FISH TAIL		Trawl	The forked end of wing of a wing trawl formed by the junction of top and lower toes. *Syn* swallow tail (NE Scot); V-wing (Grimsby)	12
Fish trap	W Scot	Trawl	FLAPPER	
FISHING LINE		Trawl Danish seine	The main frame rope along the leading edge of the lower panel. *Syn* belly cord (beam trawl, SW Scot); fish line (N Ire); jackstay (NE Eng); lower rope (Danish seine, Moray Firth); sole rope (common); under heading (S Eng)	9
Fishing line	Sussex	Trawl	WING LINE	

Term	Locality	Gear Type	Definition	Fig
FISHING LINE LEG		Trawl	Wire connecting fishing line to dan leno	17
Fishing square	Aberdeen Moray Firth	Trawl	SQUARE	
FLAG		Various	Piece of bunting surmounting dan to aid visual location	5 7 30 31
Flag	Scot	Danish seine	DAN	
Flag end	SE Scot	Pots	DAN	
Flaggy bow	NE Eng	Longline Pots	DAN	
Flap	NE Eng	Trawl	FLAPPER	
FLAPPER		Trawl	Tapered net section fitted inside extension piece or codend forming a non return fish valve. *Syn* fish trap (W Scot); flap (NE Eng); floppa (Grimsby); flopper (Hull); trap (NE Eng); valve (S Eng)	12
Flat head shackle	NW Scot	General	SHACKLE WITH COUNTERSUNK SCREW PIN	
Flat link	Hull S Eng	Trawl	RECESSED LINK	
FLEET		Gill net Set net	Any number of nets joined end to end and operated as a complete outfit. *Syn* tear (Cornwall); thread (W Scot); tier (Cornwall); train (N Ire, W Scot)	5
Fleet	1 NE Eng 2 SE Eng	Pots Pots	STRING Number of single pots	
Fleet of lines	Moray Firth	Longline	LONGLINE	
Flies	Moray Firth SE Scot	Handline	SET OF FEATHERS	
FLOAT		Various	A buoyant unit used to give static lift or to mark the position of a gear, or both. See also BUOY, MARKER BUOY, DAN, HEADLINE FLOAT	3 4 13 27
Float line	Common	Purse seine Trawl	FLOAT ROPE	
FLOAT ROPE		Purse seine Trawl	Thin rope on which numbers of floats are mounted and then seized to headline. *Syn* bog line (Hull); can line (Grimsby); cork rope (purse seine, Moray Firth); float line (common)	
Floating trawl	Common		MIDWATER TRAWL	
Floor	NW Scot	Pound net	APRON	
Floppa	Grimsby	Trawl	FLAPPER	
Flopper	Hull	Trawl	FLAPPER	

Term	Locality	Gear Type	Definition	Fig
FLOW		Net making	The degree of fullness determined by the hanging ratio	
Flue	Dorset	Longline	Fluke of dan anchor to which dan line is sometimes attached	
Flue net	Literature	Splash net	SPLASH NET	
Flush pin shackle	Common	General	SHACKLE WITH COUNTERSUNK SCREW PIN	
Fly mask	Shetland	Net making	FLY MESH	
FLY MESH		Net making	Mesh at edge of netting with only one side common with an adjacent mesh. *Syn* fly mask (Shetland)	1
FLY NET		Pound net	Stake net with leader supported throughout by stakes implanted in the sea bed and guy ropes	
Fly net	Cornwall	Drift net	DRIFT NET, for pilchard, worked with a swing rope to the first net only and without a messenger along length of fleet	
FOLLOWER		Longline	Inflated buoy attached to dan line between dan and anchor	
FOOT		Net making	The final row of a piece of netting	
Foot	Moray Firth	Longline	MAIN LINE	
FOOT CORD		Gill net Drift net	Light line to which lower edge of lint or guarding is stapled and connected to the solerope by norsels. *Syn* braiding rope (W Scot); cord (W Scot); headstring (Moray Firth); side cord (S Scot)	4
Footrope	Common	Various	General term for lower frame rope or groundrope	
Footrope leg	Hull Grimsby	Trawl	BOTTOM LEG	
Footrope link	Aberdeen W Scot	Trawl	TOGGLE	
Footrope net	Cornwall	Drift net	DRIFT NET, with lower border strengthened for attachment of the messenger by means of strops	
Footrope wire	S Eng	Trawl	BOTTOM LEG	
Forcers	Cornwall	Pots	FILLERS	
Fore bracket	S Eng	Trawl	SMALL BRACKET	
Forelock shackle	Aberdeen	General	SHACKLE WITH FORELOCK PIN AND COTTER	
Forward angle	Granton	Trawl	SMALL BRACKET	
Forward triangle	W Scot	Trawl	SMALL BRACKET	
FOUR PANEL TRAWL			Trawl comprising four panels: top, bottom and two sides. *Syn* box trawl (Fleetwood); four seam trawl (common)	10 11

2*

Term	Locality	Gear Type	Definition	Fig
Four seam trawl	Common		FOUR PANEL TRAWL	
Fram link	E Eng	General	NORSE LINK	
FRAME		1 Dredge	The triangular steel framed fore part of a shellfish dredge	28
		2 Trawl	The U-section steel outer frame of an otter board	19
Frame chain	Devon	Trawl	CHAIN GROUNDROPE, on a beam trawl	
FRAME ROPE		Various	Main support rope attached to one of the limits of a net, *eg* headline, sole rope, gable	
FRAP		Pound net	Square board buried in ground to anchor sole rope of a kettle net	
FRAP LINE		Pound net	Line joining frap to sole rope	
French dan leno	Grimsby	Trawl	DAN LENO TRIANGLE	
FRENCH POT		Pots	Cylindrical pot constructed of wooden laths and hoops, with entry in side of cylinder, fished with entry uppermost	29
FULL MESH		Net making	A complete mesh	
FULL SET		Trawl	Set of bobbins	
Full shape	Hull	Trawl	BOBBIN, spherical steel	
FUNNEL		1 Pots	Funnel shaped passage leading to eye. *Syn* criny (whelk pot, E Eng); entrance (W Scot); guide in (W Scot); monk (NE Eng); monkey (Moray Firth); mouth (SW Eng); neck (S Eng, SW Eng); tunnel (S Eng)	29
		2 Trawl Danish seine	The tapering part of a complete net leading to the codend	
FYKE NET			Anchored net comprised of leader(s) and one or more small chambers, each with an inner conical-shaped non-return 'valve' leading to the next chamber. *Syn* eel trap (Granton, Moray Firth); trap net (SE Scot)	
G-HOOK		Trawl	Forged G-shaped hook to connect with a recessed link. *Syn* coupling hook (Grimsby); G-link (common)	21
G-HOOK AND LINKS		Trawl	A G-hook on a short chain of two or three links	21
G-HOOK ASSEMBLY		Trawl	G-hook and links connected to swivel towing chain with recessed link. *Syn* G-link assembly (Hull, Grimsby)	21
G-hook swivel and link	Aberdeen	Trawl	SWIVEL TOWING CHAIN WITH RECESSED LINK	
G-link	Common	Trawl	G-HOOK	

Term	Locality	Gear Type	Definition	Fig
G-link and swivel	Moray Firth NW Scot	Trawl	SWIVEL TOWING CHAIN WITH RECESSED LINK	
G-link assembly	Hull Grimsby	Trawl	G-HOOK ASSEMBLY	
GABLE		Various	Strengthening or constraining rope fixed to the end meshes and joined to or near the ends of the upper and lower frame ropes of many nets. *Syn* deep string (ring net, W Scot); end cord, end cording (drift net, Cornwall); end line (beach seine, N Ire); end lining (drift net, Cornwall); gale (Moray Firth, SE Scot, SW Scot); gavel (NE Scot); geval (ring net, W Scot); head (Danish seine, SE Scot); heading (drift net, Lowestoft, S Eng); lug (W Scot); lug string (SE Scot, W Scot); side cord (set net, S Eng); side line (beach seine, S Eng); up and down line (trammel net, S Eng)	
Gable	1 Scot 2 SE Scot	Trawl Pound net	WING LINE CLEEK POLE	
Gable end	SE Scot	Pound net	HEAD	
Gag	Granton SE Scot	Trawl	QUARTER ROPE	
Gag line	SE Scot	Trawl	HAULING LEG	
Gale	1 Moray Firth SW Scot SE Scot	Drift net	GABLE	
	2 Cornwall	Pots	BRAID	
	3 Moray Firth	Danish seine	WING LINE	
Gathering	W Scot	Trawl	QUARTER MESHES	
Gatling	W Scot	Longline	LONGLINE	
Gavel	1 NE Scot	Drift net	GABLE	
	2 Moray Firth	Danish seine	WING LINE	
Geer	Granton	Danish seine	DAN LENO HOOP	
Geval	W Scot	Ring net	GABLE	
GILL NET			Usually rectangular in shape, made of thin twine, which catches fish by holding them in the meshes, *eg* drift net, set gill net	3 4 5
Go-line	SE Scot	Trawl	LAZY DECKIE	
Goosefoot	Grimsby	Trawl	BRIDLES (three on beam trawl)	
Goring	W Scot	Trawl	QUARTER MESHES	
GOURDON CREEL		Pots	Robustly constructed wooden creel used mainly for fishing crabs in Scotland	29

Term	Locality	Gear Type	Definition	Fig
Graith	1 Moray Firth	Longline	TIPPING	
	2 SW Scot	Longline	SNOOD	
Grape	SW Eng	Longline	DAN ANCHOR	
GRAPNEL		Various	A three or four fluked small boat anchor used at the end of a light heaving line to retrieve the dan. *Syn* cadge (Aberdeen); killick (Cornwall)	
GRASS ROPE		Trawl Danish seine	Groundrope—usually of loosely laid fibre rope weighted along its length and attached to fishing line in bights. *Syn* bass rope (Grimsby, SE Scot); chafer (NE Scot)	14
Gratlin	NE Eng	Longline	LONGLINE	
Great line	Scot	Longline	LONGLINE	31
Grimmet	NE Scot	Pots	BUTTON	
GROMMET		Trawl	Rope ring, usually made of wire, used to attach groundrope to fishing line. *Syn* grummet (Hull); sling (Aberdeen); stop (W Scot); tie (Granton); tier (Cornwall); towie (Moray Firth)	14
Ground cable	Aberdeen NE Eng	Trawl	BRIDLE	
GROUND LINE		Set gill net Tangle net	Single rope, often weighted, between anchor and bridles. *Syn* groundrope (Cornwall)	5
Ground sheet	SE Eng	Trawl	BELLY	
GROUND WIRE	Grimsby	Trawl	Bare wire replacing both bottom leg and wing section of groundrope	
Ground wire	N Wales	Trawl	BRIDLE	
GROUNDROPE		Trawl Danish seine	Connected sections of rope, usually of wire, protected with rope rounding or rubber discs or various types of bobbins, attached to and in front of the fishing line, to shield lower leading margin of a bottom trawl from ground damage whilst maintaining ground contact. *Syn* bobbins and legs (W Scot); bosom and legs (W Scot); footrope (common); solerope (Scot); stowing rope (beam trawl, SW Scot)	14
Groundrope	1 SE Scot W Scot	Pots	BACKROPE	
	2 Cornwall	Set gill net Tangle net	GROUND LINE	
Groundrope leg	E Eng	Trawl	BOTTOM LEG	
Groundrope line	W Scot	Trawl	QUARTER ROPE	

Term	Locality	Gear Type	Definition	Fig
Groundrope spreader	Granton	Trawl	BOTTOM LEG	
Groundrope spreader wire	Granton	Trawl	BOTTOM LEG	
Grummet	Hull	Trawl	GROMMET	
Guard	Scot	Various	GUARDING	
Guide-in	W Scot	Pots	FUNNEL	
GUARDING		Various	Border of netting made of stronger twine. *Syn* big heading (drift net, Cornwall); boarder (Moray Firth); double mesh (SE Scot); guard (Scot); hoddy (drift net, Lowestoft); raw (ring net, W Scot); reining (drift net, E Eng); row (W Scot)	4 6
GUSSET		Trawl	Triangular net section inserted to strengthen or enlarge. *Syn* panel (beam trawl, Devon)	10 12
Gusset	N Ire	Danish seine	CROWN	
Gut	Cornwall	Handline	CAST	
Half bobbin	Hull	Trawl	DAN LENO SCUTTLE	
Half bowl	Lowestoft	Drift net	MARKER BUOY, of distinguishing colour, marking centre of fleet	
Half egg	Grimsby	Trawl	HALF SHAPE BOBBIN	
HALF SHAPE BOBBIN		Trawl	Half-egg-shaped bobbin, usually made of composition rubber and either solid or with fillets	15
HALFER		Net making	Product of a single bar cut leaving three bars to one knot. *Syn* halver (common); harbour (Lowestoft); heffer (Sussex); three legger (S Eng)	1
Halver	Common	Net making	HALFER	
Halvering becket	E Scot	Trawl	HALVING BECKET	
HALVING BECKET		Trawl	Stout handling rope, usually combination wire and fibre, encircling fore part of codend. *Syn* bag becket (SE Scot, NE Scot); bag strop (W Scot); choker (Moray Firth); codend becket (Moray Firth); cutoff (Devon); double bag becket (Hull); doubling becket (Hull); halvering becket (E Scot); lifting becket (Moray Firth); single bag becket (SE Scot); splitter (W Scot, N Ire); splitting becket (Hull, Scot); splitting rope (Moray Firth); splitting strop (Grimsby, Lowestoft)	22
HALVING BECKET RINGS		Trawl	Metal or plastic rings fixed around codend through which the halving becket passes. *Syn* bag becket rings (Hull)	

Term	Locality	Gear Type	Definition	Fig
Hand hole	Lowestoft	Various	Small hole in netting	
HANDLINE			A hand-held line with weighted end and hooks fished above the sea bed	
Handy link	Grimsby	Trawl	NORSE LINK	
Hanger	Moray Firth	Drift net	NORSEL	
HANGING		Net making	The mounting of netting to a frame or rope according to a specific relationship between the length of the rope or frame and length of the netting. Ref BS 4440:1974	
HANGING RATIO		Net making	The ratio of the length of rope to which netting is attached, compared with the length of the fully extended netting in the direction in which it is hung. See pick up, set in	
Hanging line	Moray Firth	Purse seine	STAPLING	
Hanging ring	Granton	Trawl	LIFTING RING	
Happens	NE Eng	Drift net	STAPLING	
HARD EYE		General	An eye spliced round a thimble	21
Harbour	Lowestoft	Net making	HALFER	
Harp shackle	NW Eng	General	BOW SHACKLE	
HAULING LEG		Trawl	Wire rope extension of halving becket joined to lazy deckie. *Syn* bag becket leg (Hull); codend gag (SE Scot); gagline (SE Scot); lazy deckie leg (Fleetwood)	22
HAULING LINE		Beach seine	One of two ropes by which the net is hauled. *Syn* warp (E Eng)	7
HAUL TOW DREDGE			Oyster dredge hauled in estuaries to anchored rowing boat	
Hawser	Devon	Longline	SNOOD	
HEAD		Pound net	End wall of fish court of a bag net or stake net. *Syn* end (W Scot); gable end (SE Scot)	33
Head	1 SE Scot SW Scot	Pots	BUOY ROPE	
	2 SE Scot	Danish seine	WING LINE	
HEAD CORD		Drift net	Light line to which the top edge of the lint or guarding is stapled and connected to headline by norsels. *Syn* braiding rope (W Scot); head tow (Moray Firth); headline (Scot); headrope (common); headlink (Moray Firth); headstring (Moray Firth); side cord (SE Scot); top cord (W Scot)	4
Head gusset	Moray Firth	Trawl	TOP WING GUSSET	
HEAD POLE		Pound net	Vertical pole rigged to support head of a bag net	33

Term	Locality	Gear Type	Definition	Fig
Head sticks	NE Scot	Pots	CROSS STICKS	
Head tow	Moray Firth	Drift net	HEAD CORD or FOOT CORD	
Head yarkin	Dorset	Beach seine	HEADLINE	
Headback	1 Moray Firth	Drift net	SOLEROPE	
	2 W Scot	Ring net	HEADLINE	
Heading	Lowestoft S Eng	Drift net	GABLE	
Headlicks	Moray Firth	Drift net	STAPLING	
HEADLINE		Various	The principal upper frame rope. *Syn* back (drift net, NE Eng); back cord (beam trawl, SW Scot); backrope (drift net, Cornwall; ring net, W Scot); cork back (drift net, N Ire); cork rope (drift net, ring net, set nets, common); head yarkin (beach seine, Dorset); headback (ring net, W Scot); headrope (trawl, common); small rope (SE Scot); top reps (purse seine, N Ire); upper heading (Sussex)	3 4 6 9 27 33
HEADLINE BECKET		Trawl	Short rope spliced at each end into the headline to form an eye through which the quarter rope passes. *Syn* leech line becket (Moray Firth); quarter becket (Aberdeen)	22
HEADLINE BRIDLE		Trawl	The top bridle of a twin or three bridle trawl	17 18
Headline eye	Granton	Trawl	SWIVEL TOW SHACKLE	
HEADLINE FLOAT		Trawl Danish seine	Hollow metal or plastic float usually spherical used principally on headlines to provide lift. *Syn* ball (Scot); bobber (S Eng); bog (Hull); bottle (Grimsby, W Scot); can (Aberdeen, Grimsby, W Scot); North Sea float (common); pellet (S Eng, N Wales)	13 27
Headline leg	Eng	Trawl	TOP LEG	
HEADLINE PIECE		Trawl	One section of a sectioned headline; the sections are seized end to end to form complete headline. *Syn* headline wire (common)	
Headline quarter	Grimsby	Trawl	TOP QUARTER	
HEADLINE SHACKLE		Trawl	Large medium-gauge D-shackle connecting top leg to headline. *Syn* spreading wire shackle (Aberdeen); VD shackle (Aberdeen)	
Headline spreader	Granton	Trawl	TOP LEG	
Headline tow	S Eng	Trawl	TOP LEG	

Term	Locality	Gear Type	Definition	Fig
Headline wire	1 Common 2 Aberdeen	Trawl Trawl	HEADLINE PIECE TOP LEG	
Headlink	Moray Firth	Drift net	HEAD CORD	
Headrope	Common	1 Drift net 2 Trawl	HEAD CORD HEADLINE	
Headstring	Moray Firth	Drift net	HEAD CORD or FOOT CORD	
HEART		1 General 2 Trawl	Fibre rope around which wire strands are laid. It thus forms the heart of a wire rope CHAIN HEART or WIRE HEART	
Heel	Moray Firth	Trawl	TRAWL HEAD	
Heffer	Sussex	Net making	HALFER	
Hemp	SE Scot	Longline	TIPPING	
Herring handline	NE Scot	Handline	DANDY	
HIDES		Trawl	Cowhides attached to underside of codend to prevent chafing of the netting	
Hitch	Cornwall	Various	Small damage hole in netting	
HITCHINGS		Net making	The mounting of selvedge meshes to rope, mesh by mesh, with twine successively hitched through the point of each mesh and round the rope	2
Hockey stick	Hull	Trawl	STIFFENER	
Hoddy	Lowestoft	Drift net	GUARDING	
Holland clip	Aberdeen	Trawl	CODEND WEDGE	
Hood	NE Scot	Trawl	SQUARE	
HOOK		Handline Longline	Bent, sharpened piece of steel wire usually with barb, for catching fish. *Syn* fish hook (common); huzack (E Scot); shuck (SE Scot); yook (NE Eng)	31 32
Hook rope	W Scot	Ring net	SWEEP	
Hook thread	SW Scot	Longline	BEATING THREAD	
Hoop	1 Scot 2 S Eng	Danish seine Pots	DAN LENO HOOP BOW	
Hoop bridle	Grimsby	Danish seine	DAN LENO HOOP	
HOOP NET		Lift net	Conical net attached to a hoop and baited to catch lobsters, prawns	
HORSLIN		Handline	Thin flax line sometimes inserted between backing line and gut or cast	
Hose	Dorset	Beach seine	BUNT	
Hummel	NE Eng	Pots	MARKER BUOY	
Hundred mesh piece	Fleetwood	Trawl	MENDING SQUARE, 100 meshes wide	

Term	Locality	Gear Type	Definition	Fig
Huzack	E Scot	Longline	HOOK	
Imp	SE Scot	Longline	TIPPING	
Independent bridle	Hull	Trawl	PENNANT	
Independent piece	Grimsby	Trawl	PENNANT	
Independent wire	Hull	Trawl	PENNANT	
Inkwell pot	N Wales	Pots	CORNISH POT	
Inner net	S Eng	Trammel net	LINT	
INNER SCALE		Pound net	In a bag net or stake net the netting partition on either side of the small door and which separates the doubling from the fish court. *Syn* bag wing (W Scot)	33
Iron	SE Eng	Trawl	BRACKET	
Iron bobbin	Grimsby	Trawl	STEEL BOBBIN	
Iron bridle	NE Eng	Trawl	DAN LENO TRIANGLE	
Iron head	SW Scot	Trawl	TRAWL HEAD	
Iron sleeve bobbin	Hull	Trawl	LANCASTER	
Iron spacer	SE Scot	Trawl	LANCASTER	
Italian towing chain	SE Scot	Trawl	TOGGLE	
Jack stay	NE Eng	Trawl	FISHING LINE	
Jig line	SE Scot	Handline	DANDY	
Jigger	1 SE Scot 2 NE Eng	Handline Handline	DANDY RIPPER	
Joining link	Granton	Trawl	RECESSED LINK	
JOINING (UP) ROUND		Net making	A row of hand braided half meshes in the T-direction often of distinguishing twine joining one section of netting to another	2
JUMPER NET		Pound net	Surface fishing staked bag net, the leader secured at its land and sea ends, its length free to rise and collapse with the tide	
Keddle net	Sussex		KETTLE NET	
KEEL		Trawl	Heavy steel bar fixed to bottom edge of an otter board. *Syn* shoe (E Eng); sole (Eng); sole plate (Hull, Grimsby)	19
Keel plate	W Scot	Trawl	CHAFING PLATE	
Keep	Hull	Trawl	STOPPER	
Keep pot	SW Eng	Pots	STORE POT	

Term	Locality	Gear Type	Definition	Fig
KELLY PAD		Trawl	Short steel block welded or bolted round the keel of an otter board for added protection against abrasion on hard ground. *Syn* clog (Hull); shoe (Granton, W Scot)	
Kelly stopper	Aberdeen	Trawl	STOPPER	
KELLY'S EYE		Trawl	8-shaped steel forging, the smaller ring for attachment to backstrop, the larger through which passes the bridle, to arrest the stopper at the fore end of the bridle. *Syn* ring (Grimsby); VD ring (Cornwall)	16
KETTLE NET		Pound net	One or a series of single compartment net compounds, often roofless, each with a leader. *Syn* keddle net (Sussex); kiddle net (Sussex)	
Kiddle net	Sussex	Pound net	KETTLE NET	
Killick	Cornwall	Various	GRAPNEL	
KITE		Trawl	A shearing device mounted on a false headline to lift true headline and/or to scare fish downward into the mouth of the net	13
Kite bosom wire	Aberdeen	Trawl	KITE CARRIER WIRE	
KITE BRIDLE		Trawl	Two-legged strop with common hard eye, a pair of which are attached to the front face of the kite. *Syn* kite sling (Aberdeen); kite strops (Aberdeen)	13
KITE CARRIER WIRE		Trawl	Central section of false headline to which kite is attached by bridles. *Syn* kite bosom wire (Aberdeen); kite span (Lowestoft)	13
Kite sling	Aberdeen	Trawl	KITE BRIDLE	
Kite span	Lowestoft	Trawl	KITE CARRIER WIRE	
Kite spreaders	Aberdeen	Trawl	KITE TAILS	
Kite spreading wires	Aberdeen	Trawl	KITE WIRES	
Kite strops	Aberdeen	Trawl	KITE BRIDLE	
Kite tail pieces	Aberdeen	Trawl	KITE TAILS	
KITE TAILS		Trawl	Two ropes connecting the trailing edge of the kite to the headline. *Syn* kite spreaders (Aberdeen); kite tail pieces (Aberdeen)	13
Kite towing wires	Aberdeen	Trawl	KITE WIRES	
KITE WIRES			Ropes forming part of false headline connecting kite carrier wire to dan lenos or otter boards. *Syn* kite spreading wires (Aberdeen); kite towing wires (Aberdeen)	13

Term	Locality	Gear Type	Definition	Fig
Knife	Cornwall	Dredge	BLADE	
Knife rings	Cornwall	Dredge	BLADE RINGS	
Laceage	W Scot	Trawl Danish seine	LESTRIDGE	
Lacehood	Hull	Trawl Danish seine	LESTRIDGE	
LACING		Net making	1 The method of joining two edges of netting by consecutively passing twine through adjacent meshes and hitching it at intervals 2 As above for inferior mending of torn netting 3 The twine used for these purposes	2
Lacing	NE Eng	Pots	FASTENING	
Lacing lines	NE Eng	Trawl	STORM LINES	
Lacing string	N Eng	Trawl	CODLINE	
LANCASTER		Trawl	Iron spacer with or without connecting chain used on groundropes. *Syn* becket bobbin (Hull); chain lancaster (Moray Firth); iron sleeve bobbin (Hull); iron spacer (SE Scot); lancaster bobbin (Aberdeen); lancaster bobbin with chain (Aberdeen); spacer lancaster (Moray Firth, SE Scot, SW Scot). See also DUMMY and SPACER	15
Lancaster bobbin	Aberdeen	Trawl	LANCASTER	
Lancaster bobbin with chain	Aberdeen	Trawl	LANCASTER	
Lant netting	Dorset	Beach seine	Small mesh netting in bunt of sand eel seines	
Lanyard	Cornwall	Drift net	BUOY ROPE	
Large angle iron	Aberdeen	Trawl	LARGE BRACKET	
LARGE BRACKET		Trawl	The larger bracket of a pair, usually the aft one. *Syn* aft bracket (Grimsby); aft triangle (W Scot); big bracket (S Wales); big triangle (Granton); eye triangle (Aberdeen); large angle iron (Aberdeen); large triangle (Aberdeen); main angle (Granton, Moray Firth). See BRACKET.	19
LARGE DOOR		Pound net	Opening between mid scales leading from the cleek to the doubling of a bag net or stake net. *Syn* fish door (W Scot); large trap (SE Scot); second door (W Scot); second entrance (SW Scot)	33
Large eye swivel	Moray Firth	General	LONG BOW SWIVEL	

Term	Locality	Gear Type	Definition	Fig
Large trap	SE Scot	Pound net	LARGE DOOR	
Large triangle	Aberdeen	Trawl	LARGE BRACKET	
Last mark	Hull	Trawl	SHOOTING MARK	
Lastridge	W Scot	Trawl	LESTRIDGE	
Latt	NE Eng	Pots	CROSS PLATE	
Lazy becket	Granton SE Scot	Trawl	PENNANT	
LAZY DECKIE (LAZY DECKY)		Trawl	Fibre rope used for hauling codend to ship's side. It may be secured to halving becket, or by two tails to both codend lestridges. The fore end is usually hitched to the headline. *Syn* dog line (SE Scot); dog tow (Moray Firth); dogrope (NE Eng, Scot); double bag becket extension (Hull); fancy line (S Eng); fish rope (SW Scot, SE Scot); go-line (SE Scot); messenger (SE Scot); poke line (Hull); pork line (Hull); recovery rope (SW Scot); salvation line (Granton); tripping line (Grimsby)	22
Lazy deckie leg	Fleetwood	Trawl	HAULING LEG	
Lazy leg	Devon	Trawl	PENNANT	
Lazy man	W Scot	Trawl	PENNANT	
Lazy meshes	Lowestoft	Drift net Gill net	DORMANT MESHES	
Lazy wing line	Devon	Trawl	PREVENTIVE	
Leach line	Scot	Trawl	QUARTER ROPE	
LEAD		1 Various	Ring or barrel shaped piece of lead threaded onto solerope or leadline. *Syn* lead ring (E Scot); lead weight (W Scot); seine lead (Moray Firth); weight (Moray Firth)	3 4 6 14 27
		2 Handline	Lead weight at end of line. *Syn* plummet (Cornwall)	32
LEAD-CORED LINE		Drift net Gill net	Line incorporating lead. *Syn* bottom (salmon drift net, NE Eng)	
Lead ring	E Scot	Various	LEAD	
Lead weight	W Scot	Various	LEAD	
Lead yarking	Dorset	Beach seine	SOLEROPE	
LEADER		Pound net Fyke net	Wall of netting, often supported by stakes, which guides fish to the entrance of a net enclosure or trap. *Syn* range (kettle net, S Eng)	33
Leader	1 Granton Moray Firth	Drift net	MESSENGER	
	2 SE Scot	Longline	DAN LINE	
	3 NW Scot W Scot	Pots	BACKROPE	

Term	Locality	Gear Type	Definition	Fig
LEADLINE		Purse seine	Light rope with lead weights threaded on at intervals, which is attached to sole rope for ballast, as distinct from a sole rope with lead weights threaded on	
Leadline	Common	Set gill net Beach seine	SOLEROPE with lead weights threaded on.	
Leadrope	Hebrides	Various	SOLEROPE	
LEATHERS		Pots	Pieces of leather or plastic securing netting to base of creels	29
LEECH	Cornwall	Drift net	Unroped lower edge of pilchard fly net. *Syn* skirt (Devon)	
Leech line	Scot	Trawl	QUARTER ROPE	
Leech line becket	Moray Firth	Trawl	HEADLINE BECKET	
Leech line chain	Aberdeen	Trawl	QUARTER SWIVEL CHAIN	
LEG		Trawl	One of the wires or chains connecting net to dan leno or bridle or otter board. *Syn* spreader (Scot, Grimsby); spreading wire (Scot)	16 17
Leg	S Eng	Net making	BAR	
LENGTH		Trawl	Measure of length of warp shot, 1 length = 25 fathoms	
LENGTHENER		Trawl	Untapered section(s) of netting either inserted between belly and batings and the codend or attached to end of codend to increase its length. *Syn* codend sleeve (Devon); lengthening piece (Grimsby); sleeve (Devon); sprat sleeve (Devon)	12
Lengthening piece	Grimsby	Trawl	LENGTHENER	
Leno	S Eng	Trawl	BRIDLE	
LESTRIDGE		Trawl Danish seine	The bulky seam formed by gathering together adjacent side margins, several meshes wide, of two panels of a net and lacing them together. *Syn* laceage (W Scot); lacehood (Hull); lastridge (W Scot); selvedge (Eng)	9
LIFT NET			Baited net often held open diagonally by two spars which is lowered to the bottom and then lifted quickly to catch eels etc	
Lifter	W Scot	Trawl	QUARTER ROPE	
LIFTING BAG		Trawl	Large mesh netting cover for a codend. *Syn* chaver (Moray Firth, NE Scot, W Scot); codend double net (Hull); cover (Aberdeen, Moray Firth, NE Scot); lifting piece (W Scot); scouger (W Scot)	
Lifting bar	W Scot SW Scot	Dredge	BAR	

Term	Locality	Gear Type	Definition	Fig
Lifting becket	Moray Firth	Trawl	HALVING BECKET	
Lifting link	Hull	Trawl	LIFTING RING	
LIFTING NORMAN	Hull	Trawl	Large U-bolt located centrally on the front face of an otter board by which to lift it to or out from the gallows. See also LIFTING RING	19
Lifting norman	Hull	Trawl	LIFTING RING	
Lifting piece	W Scot	Trawl	LIFTING BAG	
LIFTING RING		Trawl	As alternative to lifting norman, a ring bracketed centrally to face of otter board. *Syn* door ring (Aberdeen); hanging ring (Granton); lifting link (Hull); lifting norman (Hull); middle norman (Fleetwood); middle ring (common)	
Lifting rope	W Scot	Trawl	QUARTER ROPE	
LIFTING STROP	SW Scot	Dredge	Floated bridle connected to both ends of bar	28
Line	1 Common 2 Eng 3 Eng	Longline Longline Longline	LONGLINE MAIN LINE STRING	
LINE ANCHOR		Longline	Small anchor at end of each section of main line	
Line back	SE Scot	Longline	MAIN LINE	
Line end	Moray Firth	Longline	DAN LINE	
Liner thread	SE Scot	Longline	BEATING THREAD	
LINK		General	1 One steel loop or ring of a chain 2 General term for any simple device that facilitates joining two or more parts of a gear	
LINT	1 Drift net Gill net 2 Trammel net		Netting in the main body of the net. *Syn* body (SE Scot); middle twine (Cornwall); middle yarn (N Ire); netting (W Scot); sheet (SW Scot); webbing (SW Scot); yarn (E Scot) The smaller mesh panel. *Syn* inner net (S Eng)	3 4
Littling	Sussex	Net making	BATING	
Lobster pot	Cornwall	Pots	CORNISH POT	
Lock	Moray Firth	Trawl	STOPPER	
LON-END	SW Eng	Beach seine	Shorter hauling line between first end shot and the shore	
LONG BOW SWIVEL		General	Swivel with one round eye and the other much elongated. *Syn* large eye swivel (Moray Firth); oblong swivel (Aberdeen)	24
LONGLINE			A number of anchored and connected lines, each bearing a large number of	31

Term	Locality	Gear Type	Definition	Fig
			baited hooks. *Syn* bolters (Devon); boulter (Cornwall); bulties (Cornwall); fleet of lines (Moray Firth); gatling (W Scot); gratlin (NE Eng); great line (Scot); line (common); stretch (Moray Firth); tear, tier (Cornwall); train (W Scot); trot line (SW Eng)	
LONG SPLIT		Pots	of creels, non-load bearing longitudinal wooden slat between the bulls in bottom of some creels. *Syn* middle piece (NE Eng)	
Loop	W Scot	Longline	DOUBLING	
Low bag	W Scot	Ring net	LOWER BUNT	
Lower bag	SE Scot W Scot	Trawl	BELLY	
LOWER BUNT		Purse seine Ring net	The lower section of a bunt when comprised of two or three sections one above the other. *Syn* bottom nets (SE Scot, W Scot); low bag (W Scot)	6
Lower bunt	Moray Firth	Trawl	BUNT	
LOWER END	Lowestoft Devon	Trawl	Aft part of the net comprised of belly and batings, lengthener and codend, all joined together	
Lower fish tail	Moray Firth	Trawl	LOWER TOE	
Lower half	Lowestoft	Trawl	LOWER END	
Lower leg	Lowestoft	Trawl	BOTTOM LEG	
LOWER PANEL		Trawl	Comprises all the net sections of the lower (underside) part of the trawl net, ie lower wings, belly, lower extension piece	9 10 11 23
LOWER QUARTER		Trawl	The corner between lower wing and belly. *Syn* bottom gathering (W Scot); bottom goring (W Scot)	9
Lower rope	Moray Firth	Danish seine	FISHING LINE	
Lower sheet	Moray Firth	Trawl	LOWER WING	
Lower spreading wire	Aberdeen	Trawl	BOTTOM LEG	
LOWER TOE		Trawl	Toe of lower wing. *Syn* bottom fish tail (Moray Firth); lower fish tail (Moray Firth); lower toe end (Moray Firth); low toe piece (Aberdeen)	9 12
Lower toe end	Moray Firth	Trawl	LOWER TOE	
Lower toe piece	Aberdeen	Trawl	LOWER TOE	
Lower towing bridle	Moray Firth	Purse seine	LOWER WING BRIDLE	

Term	Locality	Gear Type	Definition	Fig
LOWER WING		Trawl	Net section extending forward from one side of the belly and usually joined to the adjacent top wing (two panel trawls) or adjacent side wing (four panel trawls). *Syn* bottom sheet (Moray Firth); bottom wing (Cornwall); lower sheet (Moray Firth); sheet (Moray Firth); underblade (S Eng)	9 11
LOWER WING BRIDLE		Purse seine	Hauling rope sometimes attached to wing end of solerope in place of wedge. *Syn* lower towing bridle (Moray Firth)	
Lower wing bunt	Hull	Trawl	BUNT	
LOWER WING GUSSET		Trawl	Gusset between lower wing and belly. *Syn* bosom gusset (Moray Firth); bottom goring (W Scot)	12
Lud tow	NE Eng	Pots	END ROPE	
Lug	W Scot	1 Danish seine 2 Ring net Drift net	WING LINE GABLE	
Lug string	SE Scot W Scot	Ring net	GABLE	
Main angle	Granton Moray Firth	Trawl	LARGE BRACKET	
MAIN GUYS	SE Eng	Kettle net	Main anchoring ropes	
MAIN LINE		Longline	The line of a string to which the snoods are attached. *Syn* back (NE Eng, Moray Firth); backing (Cornwall, SE Scot); backing line (Devon); foot (Moray Firth); line (Eng); line back (SE Scot)	31
Main line	SE Scot W Scot	Pots	BACKROPE	
Make	Dorset	Net making	CREASING	
Making	SE Eng	Net making	CREASING	
MARKER BUOY		Various	A buoy tethered to a part of a gear to mark its position at the surface. *Syn* ball (pots, NE Eng); dan buoy (Wales); hummel (pots, NE Eng); pennet (pots, NE Eng); pot buoy (pots, SW Eng). See also TELLING and MIDDLE FLEET	
MARLING		Net making	Method of attaching netting to a frame rope by which every mesh along the selvedge is closely attached by a marline hitch with the mounting twine	2
Mash	Cornwall	Net making	CLEAN MESHES, one of	
Mat	NE Scot	Danish seine Trawl	CROWN	

Term	Locality	Gear Type	Definition	Fig
MENDING SQUARE		Trawl	A sheet of netting of suitable dimensions and mesh size for use either entire or in parts for mending damaged net sections. *Syn* hundred mesh piece (Fleetwood); shooter (Hull, Grimsby); shooting-in piece (Hull, Grimsby); sixty mesh piece (Eng)	
MESH		Net making	One of the closed spaces bounded by twine in a piece of netting	1
MESH SIZE		Net making	Expressed as the distance between the centres of two opposite knots in the same mesh when fully extended in the N direction. Ref BS 4440:1974	
MESSENGER		1 Trawl	Wire rope with steel hook at its end used in side trawling to heave both warps into the towing block	
		2 Drift net	Heavy hauling rope connecting each of the nets of a fleet of drift nets to the vessel. *Syn* bush rope (Scot); leader (Granton, Moray Firth); spring (Moray Firth, W Scot); spring rope (N Ire); warp (Lowestoft)	5
Messenger	1 Scot	Pots	BACKROPE	
	2 Granton	Pots	BUOY ROPE	
	3 Granton W Scot	Ring net	SPRING	
	4 SE Scot	Trawl	LAZY DECKIE	
Mid bag lower	SW Scot	Trawl	BELLY	
Mid head line	W Scot	Longline	MIDDLE FLEET	
MID SCALE		Pound net	In a bag net or stake net the netting partition on either side of large door leading from cleek to doubling. *Syn* second wing (W Scot)	33
MID WING LOWER		Trawl	Section of lower wing between lower toe piece and bunt	12
MID WING SIDE		Trawl	Section of side wing between side toe and shoulder	12
MID WING TICKLER		Trawl	Tickler chain attached to groundrope halfway along lower wing	
MID WING TOP		Trawl	Section of top wing between top toe piece and shoulder	12
MIDWATER TRAWL			A trawl, usually four-panelled, designed to work clear of the bottom. *Syn* floating trawl (common); pelagic trawl (common)	11
Middle	W Scot	Trawl	BOSOM	
Middle bag	Scot	Ring net	MIDDLE BUNT	
Middle bridle	SE Scot W Scot	Trawl	MIDDLE LEG	
MIDDLE BUNT		Ring net	Middle section of the bunt when it consists of three sections viz top, middle and lower. *Syn* bottom sling (Granton); middle bag (Scot)	6

Term	Locality	Gear Type	Definition	Fig
MIDDLE BUOY		Ring net	Small marker buoy attached to centre of headline	7
Middle court	W Scot	Pound net	DOUBLING	
MIDDLE CUT OFF	Devon	Trawl	Hauling rope attached to a becket surrounding middle of sprat sleeve of a midwater trawl	
MIDDLE FLEET		Longline	Dan or marker buoy attached to longlines and an anchor to mark at the surface the middle of the longline. *Syn* mid head line (W Scot). See also TELLING	
MIDDLE LEG		Trawl	The middle rope of a three-leg system, often connecting boltrope or lestridge to dan leno or bridle. *Syn* boltrope leg (Granton); boltrope spreader (Granton); boltrope spreading wire (Aberdeen); centre wire (Aberdeen); middle bridle (SE Scot, W Scot); middle spreader (Scot); middle spreading wire (Scot); middle sweep (W Scot)	17
Middle nets	W Scot	Ring net	BUNT	
Middle norman	Fleetwood	Trawl	LIFTING RING or LIFTING NORMAN	
Middle panel	NE Eng	Trawl	SIDE PANEL	
Middle piece	NE Eng	Pots	Central joist of the bottom of a creel	
Middle pocket	Moray Firth	Pound net	DOUBLING	
Middle ring	Common	Trawl	LIFTING RING	
Middle spreader	Scot	Trawl	MIDDLE LEG	
Middle spreading wire	Scot	Trawl	MIDDLE LEG	
Middle sweep	W Scot	Trawl	MIDDLE LEG	
Middle trap	SE Scot	Pound net	DOUBLING	
Middle twine	Cornwall	Drift net	LINT	
Middle wing section	SE Scot	Trawl	SIDE PANEL	
Middle yarn	N Ire	Drift net	LINT	
Mint	W Scot	Longline	BEATING THREAD	
Money box	Hull	Trawl	CODEND, extra long. See Swag bag	
Monish	W Scot SW Scot Granton	Ring net	STAPLING	
Monk	NE Eng	Pots	FUNNEL	
Monkey	Moray Firth	Pots	FUNNEL	

Term	Locality	Gear Type	Definition	Fig
Monkey nut	Hull	Trawl	SIAMESE TWIN FLOAT	
Mooring buoy	Grimsby	Danish seine	ANCHOR BUOY	
MOUNTING		Net making	Attachment of netting to frame ropes	2
Mounting rope	1 SE Scot 2 SE Scot	Purse seine Trawl	STAPLING BOLSH	
MOUTH		Trawl Danish seine	The open end through which fish enter the net	
Mouth	SW Eng	Pots	FUNNEL	
Mouth piece	Hull	Danish seine	CROWN	
Muffling	NE Eng	Trawl	SOFT BOSOM	
Murderer	SE Scot	Handline	RIPPER	
N-DIRECTION		Net making	The direction at right angles to the general course of the netting yarn. Ref BS 4440: 1974	1
Narrow nets	W Scot	Ring net	The collective term for shoulders and bunt	
Narrow wing	W Scot	Ring net	SHOULDER	
Neck	S Eng SW Eng	Pots	FUNNEL	
NET			A fishing implement comprised mainly of netting	
Net	W Scot SW Scot	Dredge	BACK	
Net bag	W Scot	Dredge	BACK	
Net end	Granton	Trawl	CODEND	
Net pot	SW Eng	Pots	Pot with netting over a rigid frame	
Net stray	Cornwall	Tangle net	Double bridle connecting end net to groundline	
NETTING			A mesh structure of indefinite shape and size; Ref BS 4440:1974. Open material of regular mesh formation made from twisted or plaited yarn, or monofilament. *Syn* webbing (common)	1
Netting	W Scot	Drift net	LINT	
Netting top	N Wales	Dredge	BACK	
Nip	S Eng	Trawl	SOFT BOSOM	
Nipper	NE Eng	Pots	BUTTON	
NORMAN		Trawl	See BACKSTROP NORMAN	
Norsal	E Eng	Gill net	NORSEL	
NORSEL LINK		General	Oval link comprising two halves with threaded ends and two connecting lock nuts. *Syn* fram link (E Eng); handy link (Grimsby)	

Term	Locality	Gear Type	Definition	Fig
NORSEL		Gill net	Separate length of cord connecting at a distance the netting to the frame rope. *Syn* daffin (NE Scot); hanger (Moray Firth); norsal (E Eng); nossle (Sussex); orsel (Cornwall); ossel/ozel (Scot); sneed (Moray Firth); snood (Moray Firth)	2 4
North Sea float	Common	Trawl	HEADLINE FLOAT, normally 20.5 cm (8 in) diameter for depths not more than about 180m	
North wing	W Scot	Pound net	OUTER SCALE	
Nossle	Sussex	Drift net	NORSEL	
NORWEGIAN SWIVEL		Handline	Small brass swivel	
O-link	Aberdeen W Scot	Trawl	RECESSED LINK	
Oblong swivel	Aberdeen	General	LONG BOW SWIVEL	
Orsel	Cornwall	Drift net	NORSEL	
Ossel	Scot	Drift net	NORSEL	
OTTER TRAWL			A trawl held open horizontally by otter boards	8
OTTER BOARD		Trawl	Shearing device, two of which hold open horizontally the wings and mouth of a trawl. *Syn* board (SE Scot); door (common); otter door (NE Eng); trawl board (common); trawl door (common)	8 19 20
OTTER BOARD TICKLER		Trawl	Tickler chain running from one otter board to the other. *Syn* door tickler (Lowestoft); door to door tickler (Lowestoft)	
Otter door	NE Eng	Trawl	OTTER BOARD	
Outer bag	W Scot	Pound net	DOUBLING	
Outer court	W Scot	Pound net	CLEEK	
Outer net	W Scot	Trammel net	ARMOURING	
OUTER SCALE		Pound net	The net walls each side of the entrance to the cleek of a bag net or stake net. *Syn* north wing (W Scot); spur (Scot)	33
Outer wall	S Eng	Trammel net	ARMOURING	
Outwall/ outwalling	S Eng	Trammel net	ARMOURING	
Oval	W Scot	Trawl	RECESSED LINK	
OVAL OTTER BOARD		Trawl	Oval shaped otter board often having vertical slots or louvres. *Syn* BMV board (Hull)	20
Oval swivel	Aberdeen	General	SHORT BOW SWIVEL	
Overcome	S Eng	Net making	FIRST ROUND	

Term	Locality	Gear Type	Definition	Fig
Overhang	W Scot	Trawl	SQUARE	
Overing	S Eng	Net making	FIRST ROUND	
Overliers	SW Scot	Pots	CROSS STICKS	
OYSTER DREDGE			Dredge used by motor or sailing vessels for oyster fishing. Smaller versions are used from rowing boats	
Ozel	Scot	Drift net	NORSEL	
Pair of bellies	Hull Fleetwood	Trawl	BELLY AND BATINGS	
PAIR TRAWL			Midwater or bottom trawl towed and held open horizontally by two vessels	
Pallet	W Scot N Ire	Various	BUOY	
PANEL		1 Trawl Danish seine	All the joined net sections between two lestridges eg TOP PANEL, SIDE PANEL, LOWER PANEL	9 10 11
		2 Net making	Sheet of netting often comprising two or more sections joined together	
Panel	Devon	Beam trawl	GUSSET	
Paravane	Devon	Trawl	Triangular dan leno with broad flat base attached by short legs to net used in banking gear	
PARCELLED		General	Wrapped with sacking, usually of wire groundropes	
PARLOUR		Pots	Inner chamber of a parlour creel. Syn boudoir (S Eng)	
PARLOUR CREEL		Pots	Crab/lobster creel of Scottish pattern with two chambers, the first holding the bait and leading to the second chamber (the holding chamber or parlour) by way of a funnel and eye in the dividing net wall. Syn stalker (S Eng)	
Parlour monk	NE Eng	Pots	Funnel and eye in the dividing net wall of a parlour creel	
PARTING SHACKLE		Trawl	Long-sided, wide-jawed D-shackle providing a weak link especially for a weak connection of groundropes	
PATERNOSTER		Handline	Light wire rod attached to handline on which to 'outrig' a baited hook. See also DANDY	
Peak	Granton	Trawl	TOP WING	
Peanut	Hull	Trawl	SIAMESE TWIN FLOAT	
Peg iron	Cornwall	Set gill net Tangle net	SINKER	
Pelagic trawl			MIDWATER TRAWL	11

Term	Locality	Gear Type	Definition	Fig
Pellet	1 W Scot	Drift net Set net	BUOY	
	2 S Eng	Trawl	HEADLINE FLOAT, plastic	
	3 N Wales	Trawl	HEADLINE FLOAT, small plastic	
PENNANT		Trawl	Handling wire connecting warp to bridle and allowing the bridle to by-pass the otter board when shooting or hauling the gear. *Syn* independent bridle (Hull); independent piece (Grimsby); independent wire (Hull, Grimsby); lazy becket (Granton, SE Scot); lazy leg (Devon); lazy man (W Scot); pick up wire (Grimsby); recovery wire (S Eng); slack back (S Eng); sweep pennant (Aberdeen)	16
Pennant	1 W Scot	Longline	DAN LINE	
	2 W Scot	Pots	BUOY ROPE	
Pennant stopper	Aberdeen	Trawl	STOPPER	
Pennet	NE Eng	Pots	MARKER BUOY	
Peter net	E Eng		SPLASH NET	
Pi-r-squared	Grimsby	Trawl	RING LINE	
Pick up	1 Grimsby	Net making	HANGING RATIO expressed as a percentage	
	2 Grimsby Hull SW Eng	Net making	CLEAN MESHES	
PICK UP WIRE		Danish seine	Wire rope between anchor rope and anchor buoy	
Pick up wire	Grimsby	Trawl	PENNANT	
Piece	NE Eng	Longline	STRING	
PILOT ANCHOR		Danish seine	Small anchor used in addition to main anchor when fishing in strong tides	
Pin	NW Scot	Pound net	STAKE	
Pipe	W Scot	Trawl	EXTENSION PIECE	
Pitcher	Cornwall	Pots	BAIT STICK	
Plate	NW Scot	Trawl	DAN LENO WASHER	
Plummet	Cornwall	Handline	LEAD	
POCKET		Trawl	Non-return valve at aft end of belly of a two panel trawl made by lacing together belly and batings to form a funnel shaped passage. *Syn* side pocket (NE Eng)	
POINT		Net making	Corner of a mesh with two bars to one knot produced by a cut in the N-direction (*cf* CLEAN MESH). *Syn* cut knot (Dorset); cut mesh (E Eng); diamond (Hull); side knot (common)	

Term	Locality	Gear Type	Definition	Fig
Point	1 Granton 2 Hull	Trawl Trawl	TOP WING TOE	
Poke	S Eng SW Eng	Trawl	CODEND	
Poke lashing	S Eng	Trawl	CODLINE	
Poke line	1 Devon 2 Hull	Trawl Trawl	CODLINE LAZY DECKIE	
POLE		Pound net	See CLEEK POLE or HEAD POLE. *Syn* snood (Moray Firth)	
Pole	Scot	Trawl Danish seine	DAN LENO STICK	
POLE END	Lowestoft	Drift net	First net shot and hence normally last to be hauled. *Syn* tayse end (Cornwall)	
POLE BUOY	Lowestoft	Drift net Longline	Marker buoy or illuminated dan, marking far end of fleet or line. *Syn* tayse buoy (Cornwall)	
Polie	Moray Firth	Danish seine	DAN LENO	
POLLACK NET			Loosely hung set gill net with wooden pole at each end	
PONY BOARD		Trawl	Small otter board used in place of a dan leno	
Pork line	Hull	Trawl	LAZY DECKIE	
Posts	W Scot	Ring net	STAPLING	
POT			Small trap, baited with fish or offal etc, with one or more conical or funnel shaped entrances. Often qualified by name of species it is intended to catch *eg* crab pot, lobster pot, or area in which a particular design is used, *eg* Cornish pot, Scottish creel, or the material of which it is made, *eg* withy pot, wire pot. See CREEL	29
POT BRIDLE		Pots	Loop of rope attached to pot for attachment to pot strop. *Syn* sides (S Eng); underbridle (NE Eng)	29 30
Pot buoy	SW Eng	Pots	MARKER BUOY	
Pot leg	Cornwall	Pots	POT STROP	
POT LINE		Pots	10–12 fathom section of the backrope. *Syn* pot rope (S Eng); pot tow (S Eng, SW Eng); tow (Wales); trot rope (S Eng)	30
Pot line	S Eng	Pots	POT STROP	
Pot rope	S Eng	Pots	POT LINE	
POT SPINNER		Pots	Small 8-shaped steel link for connecting pot bridle to pot strop	
POT STONE		Pots	Flat stone lashed to bottom of pot for ballast. *Syn* side stone (Cornwall)	29

Term	Locality	Gear Type	Definition	Fig
POT STROP		Pots	Branch line joining backrope either to the pot bridle or directly to the pot. *Syn* daphnes (NW Scot); dropper (NW Scot); pot leg (Cornwall); pot line (S Eng); tail (W Scot)	30
Pot tow	S Eng SW Eng	Pots	POT LINE	
POUND NET			General term for any moored and/or staked net comprising a leader and one or more enclosures, *eg* bag net, stake net, kettle net	
Preen	Cornwall	Pots	BAIT STICK	
PREVENTIVE		Trawl Danish seine	Rope or chain connected to headline and fishing line at or near their ends for extra strength. *Syn* lazy wing line (Devon); save-all (Lowestoft); tying down line (S Eng)	
Preventive	Granton	Danish seine	WING LINE	
Preventive bar	Aberdeen	Trawl	STIFFENER	
Prime	Cornwall	Pots	BAIT STICK	
Protective plate	Aberdeen	Trawl	DAN LENO WASHER	
Protector	Aberdeen	Danish seine	WING LINE	
Purse floats	SE Scot	Purse seine	PURSE SEINE FLOATS	
Purse hook	Moray Firth	Purse seine	PURSING RING	
PURSE SEINE			A large single-panel multi-sectioned net used to encircle pelagic fish, the bottom of which is then drawn together to enclose them	6 7
PURSE SEINE FLOATS		Purse seine	Oval shaped solid plastic floats used mainly on purse seine to support the headline. *Syn* purse floats (SE Scot)	7
PURSING RING		Purse seine	Strong metal clip ring with spring loaded gate through which the pursing wire passes. *Syn* boss hook (Moray Firth); purse hook (Moray Firth); skagen hook (Aberdeen)	7
PURSING RING BECKET		Purse seine	Rope connecting pursing ring to pursing ring bridle	7
PURSING RING BRIDLE		Purse seine	Rope connecting pursing ring becket to two points on the sole rope. Often refers to the general rope arrangement that connects the pursing ring to the sole rope. *Syn* pursing strop (Moray Firth)	7
Pursing strop	Moray Firth	Purse seine	PURSING RING BRIDLE	
PURSING WIRE		Purse seine	Steel wire running through the pursing rings by means of which the bottom of the net is closed	7

Term	Locality	Gear Type	Definition	Fig
PUSH NET			A hand held net on a rigid frame which is pushed along the sea bed in shallow water for shrimps etc	
Pye eye	Hull	Trawl	RING LINE	
QUARTER		Trawl	Corner formed by junction of square and inner edge of top wing or top wing gusset (top quarter), or belly and inner edge of lower wing or lower wing gusset (lower quarter). *Syn* corner (Moray Firth)	9
Quarter becket	Aberdeen	Trawl	HEADLINE BECKET	
Quarter belly line	Aberdeen	Trawl	BELLY LINE	
Quarter chain	Grimsby	Trawl	QUARTER SWIVEL CHAIN	
Quarter line	W Scot	Trawl	QUARTER ROPE	
Quarter links	Grimsby	Trawl	QUARTER SWIVEL CHAIN	
QUARTER MESHES		Trawl	The few meshes taken together into one staple at the quarter. *Syn* crouping (W Scot); gathering (W Scot); goring (W Scot)	
Quarter ring and strop	Moray Firth	Trawl	QUARTER STROP	
QUARTER ROPE		Trawl	Handling rope used in side trawling to bring the bosom section of the groundrope to the ship's side. *Syn* bobbin lifting rope (W Scot); bobbin rope (W Scot, SW Scot); bosom-gag (SE Scot); gag (Granton, SE Scot); groundrope line (W Scot); leach line (Scot); leech line (Scot); lifter (W Scot); lifting rope (W Scot); quarter line (W Scot)	22
QUARTER STROP		Trawl	Wire strop spliced to steel ring used to connect quarter rope to quarter swivel. *Syn* quarter ring and strop (Moray Firth); saver (SE Scot)	22
QUARTER SWIVEL CHAIN		Trawl	Chain and swivel connecting quarter rope, or quarter strop, to ground rope. *Syn* leech line chain (Aberdeen); quarter chain (Grimsby); quarter links (Grimsby); quarter swivel and links (Granton, Grimsby)	22
Quarter swivel and links	Grimsby Granton	Trawl	QUARTER SWIVEL CHAIN	
QUEEN DREDGE			Large-framed small mesh dredge for catching queens	
Queen trap	Lowestoft S Eng	Trawl	SHELL TRAP	
Quidling	S Eng	Pots	Taut horizontal piece of netting surrounding entrance to a whelk pot to deter escape	

Term	Locality	Gear Type	Definition	Fig
Rails	NE Eng	Pots	CROSS STICKS	
Range	S Eng	Kettle net	LEADER	
Raw	W Scot	Ring net	GUARDING	
RAY NET		Tangle net	General purpose large mesh tangle net for shellfish, rays, angler fish etc	
RECESSED LINK		Trawl	Forged oval link with recesses to permit engagement with a G-hook. *Syn* flat link (Hull, S Eng); joining link (Granton); O-link (Aberdeen, W. Scot); oval (W Scot); retaining link (SE Scot); split link (Scot)	21
Recessed shackle	Grimsby	Trawl	D-SHACKLE WITH COUNTERSUNK SCREW PIN	
Recovery rope	SW Scot	Trawl	LAZY DECKIE	
Recovery wire	S Eng	Trawl	PENNANT	
Reining	E Eng	Drift net	GUARDING	
Retaining link	SE Scot	Trawl	RECESSED LINK	
Retro-rocket	Cornwall	Trawl	Hollow bullet-shaped metal casting containing a swivel that is used as a dan leno or bumper bobbin. *Syn* torpedo (Cornwall)	
Rib line	Hull	Trawl	BELLY LINE	
RIB		Pots	Vertical frame component of a pot. *Syn* stay (Devon); stud (Devon); wire (Cornwall)	29
RIG		Various	The way the components of fishing gear are assembled	
Ring	1 Grimsby	Trawl	KELLY'S EYE	
	2 W Scot	Trawl	TOGGLE	
RING BOLT		Trawl	Steel ring through a single bolt often used on small otter boards for attachment of back straps. *Syn* eye bolt (common)	
RING LINE		Trawl	Hauling rope surrounding belly and batings sometimes used in side trawling. *Syn* pi-r-squared (Grimsby); pye eye (Hull)	
RING NET			A single-panelled multi-sectioned pelagic encircling net usually operated by two vessels	6
Ringing	Cornwall	Pots	BRAID	
RINGLES	SE Eng	Dredge	The steel rings used in the chain belly of a shellfish dredge	
Rings	Moray Firth	Dredge	CHAIN BELLY	
RIPPER		Handline	A bright metal plummet with two or three pairs of hooks attached that is jigged up and down to attract and foul-hook fish. *Syn* jigger (NE Eng); murderer (SE Scot)	32

Term	Locality	Gear Type	Definition	Fig
Rolled groundrope	S Eng	Trawl	ROPE ROUNDED GROUNDROPE	
ROOF		Pound net	Section of netting that covers one or more chambers of bag net and stake net. *Syn* cover (NW Scot)	33
ROPE			Stout line made by twisting or braiding together a number of strands of wire or fibre	
Rope	Scot	Danish seine	SEINE ROPE	
ROPE ROUNDED GROUND-ROPE		Trawl	Wire or chain groundrope with fibre rope firmly twisted round it in place of bobbins, rubber discs, etc. *Syn* rolled groundrope (S Eng); rounded groundrope (Grimsby); snaked footrope (Devon); wolded groundrope (Fleetwood, Lowestoft). See also SOFT BOSOM	14
Ross bobbin shackle	Aberdeen	Trawl	V-SHACKLE used on dan leno spindle	
Ross shackle	Aberdeen	Trawl	V-SHACKLE used on dan leno spindle	
Round	Common	Net making	ROW	
Round dan leno	SE Scot	Danish seine	DAN LENO HOOP	
Rounded ground rope	Grimsby	Trawl	ROPE ROUNDED GROUNDROPE	
ROW		Net making	A row of half meshes in the T-direction. *Syn* round (common); went (Dorset)	1
Row	W Scot	Ring net	GUARDING	
RUBBER		Handline	Coloured rubber or plastic lure, usually tubular, on shank of hook	
RUBBER BOBBIN	Common	Trawl	Bobbin made of rubber or composition rubber. *Syn* Teal bobbin (common)	15
RUBBER DISCS			Discs with hole in centre, usually cut from old lorry tyres and used to clad groundrope sections. *Syn* rubbers (common)	14
Rubber footrope	Common	Trawl	RUBBER GROUNDROPE	
RUBBER GROUND-ROPE	Common		Groundrope with all sections clad with rubber discs and connecting links to fishing lines. *Syn* rubber footrope (common); rubbers (common)	
Rubber legs	Scot	Trawl	WING RUBBERS	
Rubber roller	SE Scot	Trawl	CYLINDRICAL RUBBER BOBBIN	
RUBBER SANDEEL		Handline	See RUBBER	
RUBBER SPACER		Trawl	Spacer made of rubber or composition rubber. *Syn* barrel spacer (SE Scot, Moray Firth)	14 15

Term	Locality	Gear Type	Definition	Fig
Rubber wheel	Common	Trawl	CYLINDRICAL RUBBER BOBBIN	
Rubbers	Common	Trawl	1 RUBBER DISCS 2 RUBBER GROUNDROPE 3 WING RUBBERS	
RUBBY DUBBY BAG		Line	Sack containing chopped fish, or offal, hung overboard to attract sharks etc	
Run	N Ire	Drift net	STAPLING	
Runner	1 Granton 2 SE Scot	Trawl Pots	TRAWL HEAD BULL	
Sally	S Eng	Trawl	Large shackle, used to connect pennant to bridle and serve as a stopper at the kelly's eye	
SALVAGEE	Cornwall	Drift net	Bundle of light ropes acting as a shock absorber between first net and swing rope of a pilchard fly net	
Salvation line	Granton	Trawl	LAZY DECKIE	
Sand plate	Hull Grimsby	Trawl	CHAFING PLATE	
Sandwich plate	W Scot	Trawl	DIVISIONAL BAR	
Sausage	Grimsby	Trawl	A tube of netting filled with several floats for attachment to headline of a midwater trawl	
Save-all	Lowestoft	Trawl	PREVENTIVE	
Saver	SE Scot	Trawl	QUARTER STROP	
SCALE		Pound net	See OUTER SCALE, MID SCALE, INNER SCALE	
SCALLOP DREDGE			Large heavy-framed dredge for catching scallops	28
Scawbs	NE Eng	Pots	CROSS STICKS	
Sconking	S Eng	Drift net Kettle net	Old netting wrapped around lower frame rope to improve bottom contact or give protection	
SCORE		Drift net Ring net	Measure of depth of net; one score equals 20 full meshes	
Scottish creel	Eng	Pots	CREEL	
Scouger	W Scot	Trawl	LIFTING BAG	
Screw link	Aberdeen	Trawl	CARBINE HOOK WITH SCREW KEEP	
Screwing shackle	Aberdeen	Trawl	D-SHACKLE WITH COUNTERSUNK SCREW PIN	
Scringe net	W Scot SW Scot		BEACH SEINE	
Scrub plate	Hull	Trawl	CHAFING PLATE	
SCUM NET		Drift net	A net rigged along side of vessel to catch	

Term	Locality	Gear Type	Definition	Fig
			fish that fall back from drift nets whilst hauling. *Syn* thief net (Lowestoft)	
SCUMMER		Purse seine	Hand held brailer	
SCUNNINGS		Drift net	Adjacent gables of pilchard nets joined by a number of short cords. See TACHINGS	
Scuttle	Grimsby	Trawl	DAN LENO SCUTTLE	
SCUTTLE DAN LENO		Trawl	Dan leno assembly with a scuttle bobbin	16
Second door	W Scot	Pound net	LARGE DOOR	
Second entrance	SW Scot	Pound net	LARGE DOOR	
SECOND POUND	S Eng	Pound net	The bight of a kettle net furthest from the shore	
Second wing	W Scot	Pound net	MID SCALE	
SECTION		Net making	1 Piece of netting of uniform mesh size and twine thickness 2 A principal part of a panel	1 6, 9, 10, 11, 23
SEINE			An encircling net, sometimes with ropes, *eg* PURSE SEINE, BEACH SEINE, DANISH SEINE	6 7 26 27
Seine lead	Moray Firth	Danish seine	LEAD	
Seine link	Scot	Danish seine	CLIP LINK used to connect seine rope to dan leno and sometimes to join two coils of seine rope.	
Seine net rope	Granton	Danish seine	SEINE ROPE	
Seine net swivel	Aberdeen	Danish seine	SWIVEL WITH CLIP LINK	
SEINE ROPE		Danish seine	One of the two very long ropes to haul the net, comprising a number of 120 fathom coils joined end to end. *Syn* rope (Scot); seine net rope (Granton); thread (W Scot); warp (Moray Firth, W Scot)	26
SEIZED		Net making	Tightly bound together with cordage	
Seizing	Lowestoft	Drift net	STOPPER	
Seizing(s)	Lowestoft	Trawl	SETTING(S)	
Selvage rope	N Wales	Trawl	BOLTROPE	
SELVEDGE		Net making	The edge of a piece of netting	
Selvedge	Eng	Trawl Danish seine	LESTRIDGE	
Selvedge rope	1 SE Scot 2 Granton N Wales	Purse seine Trawl	STAPLING BOLTROPE	
SERVED		General	Wrapped closely with twine, *eg* round a splice. See also SERVED HEADLINE	

Term	Locality	Gear Type	Definition	Fig
SERVED HEADLINE		Trawl	Headline wire tightly bound with cord along its entire length	
SET GILL NET			Gill net anchored to sea bed. *Syn* anchor net (Moray Firth); straight net (Dorset)	
Set in	Lowestoft	Net making	HANGING RATIO expressed as the fraction of the fully stretched dimension of the netting that is taken in when it is mounted to the frame rope, *eg* set in by the half, third etc	5
SET NET			General term for any simple net when it is held in fishing trim by anchors, sinkers and/or stakes, *eg* trammel net, tangle net, gill net	
SET OF BOBBINS		Trawl	Bobbins and lancasters threaded on bobbin wires	14
SET OF FEATHERS		Handline	A cast bearing several droppers each with a hook dressed with coloured feathers. *Syn* darrow (W Scot); feathered serpent (SE Scot); feathers (Cornwall); flies (Moray Firth, SE Scot); stand of flies (NE Scot)	32
SET OF GEAR	Grimsby Hull	Trawl	All the gear except the otter boards and warps	
Set of headropes	Cornwall	Drift net	DOUBLE HEADLINE	
SET OF RUBBERS		Handline	A cast bearing several droppers each with a hook, with strips of rubber or plastic attached	
Setting	Moray Firth	Trawl	STAPLING	
SETTING(S)		Various	Short length(s) of twine used to seize two parts of a gear together. *Syn* seizing(s) (Lowestoft); sitting(s) (Lowestoft); stopping(s) (S Eng); stop(s) (S Eng)	2
Setting line	SE Scot W Scot	Drift net	STAPLING	
SHACKLE		General	A staple-like link closed with one of a variety of pins, *eg* 1 SQUARE HEADED SCREW PIN. *Syn* spanner shackle (Moray Firth); square head shackle (Aberdeen) 2 SCREW PIN WITH EYE 3 COUNTERSUNK SCREW PIN. *Syn* flat head shackle (NW Scot); flush pin shackle (common); recessed shackle (Grimsby); screwing shackle (Aberdeen) 4 FORELOCK PIN AND COTTER. *Syn* forelock shackle (Aberdeen); wedge shackle (Aberdeen) See also D-SHACKLE, BOW SHACKLE	24
SHANK NET			Small meshed net attached to a semi-circular wooden frame, horse or tractor drawn in shallow water	

Term	Locality	Gear Type	Definition	Fig
Shearboard link	Grimsby	Trawl	BACKSTROP LINK	
Sheet	1 Moray Firth	Trawl	LOWER WING	
	2 W Scot	Trawl	BELLY	
	3 Granton	Ring net	DEEPENING	
	4 SW Scot	Drift net	LINT	
SHELL TRAP		Trawl	Section of large mesh netting in the belly which allows stones and rubbish to fall through. *Syn* queen trap (Lowestoft, S Eng); stone trap (Devon, Cornwall)	
SHERINGHAM POT			Small iron-ribbed rope-bound inkwell-shaped pot for catching whelks	
SHIP-END		Beach seine	The hauling line between the offshore wing and the beach. It is usually longer than the shore wing hauling line. See LON-END	
Shoe	1 E Eng	Trawl	KEEL	
	2 Granton W Scot	Trawl	KELLY PAD	
	3 Granton	Trawl	TRAWL HEAD	
	4 E Eng	Trawl	SOLE PLATE	
Shoe plate	Aberdeen	Trawl	CHAFING PLATE	
Shoeing plate	Aberdeen	Trawl	CHAFING PLATE	
Shooter	Hull Grimsby	Trawl	MENDING SQUARE	
SHOOTING MARK		Trawl	Warp mark nearest and at specified distance from otter board end of warp. *Syn* last mark (Hull); warning mark (Hull)	
Shooting-in piece	Grimsby Hull	Trawl	MENDING SQUARE	
Shore seine	S Eng		BEACH SEINE	
SHORT BOW SWIVEL			Swivel with one round eye and the other slightly elongated. *Syn* oval swivel (Aberdeen); small eye swivel (Moray Firth)	24
SHOULDER		1 Danish seine Ring net	Net section between bunt, or bag, and wing. *Syn* narrow wing (ring net, W Scot)	27
		2 Trawl	Rear section of top wing in front of the square	10 12
Shoulder	1 Aberdeen	Trawl	SIDE PANEL	
	2 Moray Firth NE Scot SW Scot	Trawl	BUNT	
SHOULDER BUOY		Ring net	Marker buoy attached to headline to mark junction of shoulder and top bunt	
Shrink	Dorset	Net making	BATING	

Term	Locality	Gear Type	Definition	Fig
Shuck	SE Scot	Longline Handline	HOOK	
Siamese float	Hull	Trawl	SIAMESE TWIN FLOAT	
SIAMESE TWIN FLOAT		Trawl	Headline lifter comprising two hollow spherical floats attached under a cambered canopy. *Syn* monkey nut (Hull); peanut (Hull); Siamese float (Hull)	13
SIDE		Pound net	Netting side wall of cleek, doubling and fish court in a bag net or stake net	33
SIDE BELLY		Trawl	Net section(s) between side wings and codend of a four panel trawl	11
Side belly	Grimsby	Trawl	SIDE PANEL	
SIDE BOSOM PIECE		Trawl	Narrow section of stronger netting across forward edge of side belly adjacent to wing line of a four panel trawl	12
Side cord	1 SE Scot 2 S Eng	Drift net Set gill net	HEAD CORD or FOOT CORD GABLE	
Side heading	Sussex	Trawl	WING LINE	
Side knot	Common	Net making	POINT	
Side line	1 S Eng 2 Sussex 3 Devon	Beach seine Trawl Trawl	GABLE BOLT ROPE WING LINE	
SIDE PANEL		Trawl	Comprises all the side net sections of a four panel trawl joined together. *Syn* box (W Scot); middle panel (NE Eng); middle wing section (SE Scot); shoulder (Aberdeen); side belly (Grimsby); V-piece (Devon)	10 11 23
Side panel lower wing	Hull	Trawl	SIDE WING LOWER	
Side panel top wing	Hull	Trawl	SIDE WING TOP	
Side piece	NE Eng	Pots	BULL	
Side pocket	Devon	Trawl	POCKET	
Side sticks	SE Scot SW Scot	Pots	CROSS STICKS	
Side stone	Cornwall	Pots	POT STONE	
SIDE WEIGHTS		Trawl	Additional steel plates or bars welded to keel of an otter board for extra ballast	
SIDE WING LOWER		Trawl	Lower wing of side panel of a four panel trawl. *Syn* side panel lower wing (Hull)	11
SIDE WING TOP		Trawl	Upper wing of side panel of a four panel trawl. *Syn* side panel top wing (Hull)	11
Sides	S Eng	Pots	POT BRIDLE	

Term	Locality	Gear Type	Definition	Fig
Single bag becket	SE Scot	Trawl	HALVING BECKET	
Single belly	Fleetwood	Trawl	Either belly or batings when these are identical	
SINKER		Set net Pots Longline	Concrete block, heavy stone etc used to anchor the gear. *Syn* peg iron (set nets, Cornwall); slinger (longline, S Eng); weight (W Scot)	5
Sittings	Lowestoft	Trawl	SETTINGS	
Sixty mesh piece	Eng	Trawl	MENDING SQUARE, 60 meshes wide	
SKAGEN OTTER BOARD		Trawl	Heavily built otter board with broad keel	
Skagen hook	Aberdeen	Purse seine	PURSING RING	
Skein hole	Devon	Various	Small tear in a net	
Skewer	Cornwall	Pots	BAIT STICK	
Skirt	Cornwall	Drift net	LEECH	
Skiver	Cornwall	Pots	BAIT STICK	
Slack back	1 S Eng 2 Devon	Trawl Trawl	PENNANT BRIDLE	
SLATS		Pots	Roofing laths used to construct a lobster pot	
Sledge	SE Scot	Trawl	TRAWL HEAD	
Sleepy meshes	Literature	Gill net	DORMANT MESHES	
Sleeve	Devon	Trawl	LENGTHENER	
Sliding knot	W Scot	Pots	BUTTON	
Sling	1 Granton SW Scot 2 Aberdeen 3 S Eng	Ring net Trawl Trawl	BUNT GROMMET BACKSTROP	
Slinger	S Eng	Longline	Lead sinker at end of each string of main line	
Slip knot	Moray Firth	Pots	BUTTON	
Small angle iron	Aberdeen	Trawl	SMALL BRACKET	
SMALL BRACKET		Trawl	The smaller bracket and usually the forward one on an otter board. *Syn* fore-bracket (S Eng); forward angle (Granton); forward triangle (W Scot); small angle iron (Aberdeen); small triangle (Aberdeen, Granton); twisted bracket (S Scot)	19
SMALL DOOR		Pound net	Entrance to fish court of bag net or stake	33

(cont)

Term	Locality	Gear Type	Definition	Fig
			net. *Syn* fish door (W Scot); small trap (SE Scot); third door (W Scot); third entrance (SW Scot)	
Small eye swivel	Moray Firth		SHORT BOW SWIVEL	
SMALL MESH COVER		Trawl	Small mesh sack fitted over codend to retain small fish. *Syn* cover (Aberdeen, Moray Firth, NE Scot)	
Small rope	SE Scot	Trawl	HEADLINE	
Small trap	SE Scot	Pound net	SMALL DOOR	
Small triangle	Aberdeen Granton	Trawl	SMALL BRACKET	
Snaked footrope	Devon	Trawl	ROPE ROUNDED GROUNDROPE	
Sneed	1 Scot 2 Moray Firth	Longline Drift net	SNOOD NORSEL	
SNOOD		Longline	One of the hook-carrying branch lines. *Syn* graith (SW Scot); hawser (Devon); sneed (Scot); snud (Lowestoft); strood (SE Scot); strop (Cornwall)	31
Snood	1 Moray Firth 2 Moray Firth	Pound net Drift net	POLE NORSEL	
SNOOD CLIP		Longline	Wire clip of various patterns used to attach snood to main line	
Snud	Lowestoft	Longline	SNOOD	
SOFT BOSOM		Trawl	Bosom section of groundrope rounded with rope. *Syn* muffling (NE Eng); nip (S Eng)	14
SOFT EYE		General	An eye splice without thimble	
Sole	Eng	Trawl	KEEL	
Sole bolshline	W Scot	Purse seine	STAPLING, to sole rope	
Sole line	S Eng	Drift net	SOLEROPE	
SOLE PLATE		Trawl	Flat steel plate welded to bottom of trawl head of a beam trawl. *Syn* shoe (E Eng)	
Sole plate	Hull Grimsby	Trawl	KEEL	
SOLEROPE		Gill net Beach seine	Lower frame rope, often with leads on. *Syn* bottom doubling (W Scot); head back (Moray Firth); leadline (common); leadrope (Hebrides); lead yarking (Dorset); sole line (S Eng); suckrope (drift net, Moray Firth)	3 4 6
Solerope	Scot	Various	General term for lower frame rope, *eg* fishing line and some groundropes	

Term	Locality	Gear Type	Definition	Fig
SPACER		Trawl	Small diameter bobbin inserted between principal bobbins of a groundrope. See also LANCASTER, DUMMY	14
Spacer lancaster	Moray Firth SE Scot SW Scot	Trawl	LANCASTER	
Spake	SE Scot	Pots	CROSS PLATE	
Spanner shackle	Moray Firth	General	D-SHACKLE WITH SQUARE HEAD PIN	
Spend tow	SW Scot	Longline	TROT TOW	
Spiller	Cornwall	Longline	Lightweight longline for whiting etc	
Spindle	Common	Trawl	DAN LENO SPINDLE	
Spindle ring	Aberdeen	Trawl	DAN LENO WASHER	
SPINNER		Handline	Bright metal lure that spins when trolled	
SPLASH NET			An inclined sheet of netting into or on to which fish in shallow water are frightened by splashing the water. *Syn* beating net (lit); flue net (lit); peter net (E Eng)	
Split codend	Grimsby	Trawl	DOUBLE CODEND	
Split connecting link	Aberdeen	Trawl	FALSE LINK	
Split link	Scot	Trawl	variously: FALSE LINK, RECESSED LINK, CLIP LINK	
Splitter	W Scot N Ire	Trawl	HALVING BECKET	
Splitting becket	Hull, Scot	Trawl	HALVING BECKET	
Splitting rope	Moray Firth	Trawl	HALVING BECKET	
Splitting strop	Grimsby Lowestoft	Trawl	HALVING BECKET	
Sprat sleeve	Devon	Trawl	LENGTHENER (small meshed)	
SPREADER		Handline	Single metal rod, usually of brass, with lead weight about its centre and bearing a baited hook at each end	
Spreader	1 Scot Grimsby	Trawl	LEG	
	2 Moray Firth	Danish seine	SWEEP	
	3 W Scot Cornwall	Beach seine	DAN LENO STICK	
Spreader bar	Aberdeen	Trawl	BUTTERFLY	
Spreading wire	Scot	Trawl	LEG	
Spreading wire shackle	Aberdeen	Trawl	HEADLINE SHACKLE	
Spreadline	Devon	Longline	END ROPE	

Term	Locality	Gear Type	Definition	Fig
SPRING	Scot	Ring net	Hauling line connected to sole rope by stoppers. *Syn* messenger (Granton, W Scot); spring rope (Granton, W Scot)	7
Spring	Moray Firth W Scot	Drift net	MESSENGER	
SPRING LOADED TEETH BAR		Dredge	Teeth bar with shock absorbing springs	
Spring rope	1 Granton W Scot	Ring net	SPRING	
	2 N Ire	Drift net	MESSENGER	
SPROOL	NE Scot	Handline	Two rippers attached one at each end of a spreader	32
Sprool	W Scot	Handline	DANDY	
Sprule	W Scot	Handline	DANDY	
Sprunk	Lowestoft	Various	Small tear in netting	
Spur	Scot	Pound net	OUTER SCALE	
SQUARE		Trawl	Section of top panel between headline and batings. *Syn* cover (SW Scot); dead net (Sussex); fishing square (Aberdeen, Moray Firth); hood (NE Scot); overhang (W Scot)	9 11 23
Square head shackle	Aberdeen	General	D-SHACKLE WITH SQUARE HEAD SCREW PIN	
Square lines	NE Eng	Trawl	STORM LINES	
Squinches	Cornwall	Drift net	TACHINGS	
STAKE		Pound net Stop net	Pole used to anchor and support a net. *Syn* pin (NW Scot)	
STAKE NET		Pound net	Net, similar to a bag net, supported by stakes	
Stalker	S Eng	Pots	PARLOUR CREEL, wooden	
Stand of flies	NE Scot	Handline	SET OF FEATHERS	
Staple	Common	Net making	STAPLING	
STAPLING		Net making	The loops of cord formed when netting is mounted to a frame rope by reeving the cord through the selvedge meshes and hitching at regular mesh intervals. *Syn* bights (ring net, SE Scot); bolsh rope (purse seine, W Scot); bools (ring net, Scot); bulls (W Scot); hanging line (purse seine, Moray Firth); happens (drift net, NE Eng); headlicks (drift net, Moray Firth); monish (W Scot, SW Scot, Granton); mounting rope (purse seine, SE Scot); posts (ring net, W Scot); run (N Ire); selvedge rope (SE Scot); setting (trawl, Moray Firth); setting line (drift net, SE Scot, W Scot); sole bolsh line (to solerope	2

Term	Locality	Gear Type	Definition	Fig
			of purse seine, W Scot); staple (common); striking up line (S Eng)	
Stay	Devon	Pots	RIB	
STEEL BOBBIN		Trawl	Hollow, spherical or elliptical steel bobbin. *Syn* iron bobbin (Grimsby)	15
Steel door	Scot	Trawl	V-BOARD	
Steel strap	Granton	Trawl	STIFFENER	
Stem net	Lowestoft	Drift net	BOAT NET	
Stick	Common	Trawl Danish seine Beach seine	DAN LENO STICK	
Stick bridle	Grimsby	Trawl Danish seine Beach seine	DANE LENO STROP	27
STICK CHAIN	NE Eng	Trawl	A length of chain inserted between end of bridle and dan leno strop	
Stick dan leno	Common	Danish seine Trawl	DAN LENO STICK	
Sticky bow	NE Eng	Longline Pots	DAN	
STIFFENER		Trawl	Steel bar shaped to fit round the top corner of an otter board and extend each side at an angle to the chafing plates and bolted to the planks. *Syn* end strap (Aberdeen); hockey stick (Hull); preventive bar (Aberdeen); steel strap (Granton); strap (S Eng); strapping (S Eng); strengthener (NE Eng)	19
Stolen mesh	Cornwall	Net making	CREASING	
STONE MAT		Trawl	A criss-cross of chains hung from beam of a beam trawl and connected to groundrope to keep boulders out of the net and also disturb flat fish. *Syn* chain belly (Devon); chain mat (Devon)	
Stone trap	Devon Cornwall	Trawl	SHELL TRAP	
Stop(s)	1 W Scot 2 S Eng	Trawl Trawl	GROMMET See Stopping(s)	
Stop link	Devon	Trawl	STOPPER	
STOP NET			A wall of netting set on stakes to enclose a small bay or creek at high tide	
STOPPER		1 Trawl	Forged double eyed link with shoulders connecting pennant to bridle. *Syn* cable keep (Hull); figure of eight link (Grimsby); keep (Hull); kelly stopper (Aberdeen); lock (Moray Firth); pennant stopper (Aberdeen); stop link (Devon); stopper *(cont)*	16

Term	Locality	Gear Type	Definition	Fig
			link (Fleetwood, Hull); sweep stopper (Aberdeen)	
		2 Drift net	Rope connecting ends of adjacent sole ropes to the messenger. *Syn* seizing (Lowestoft)	5
		3 Ring net	One of a number of rope strops connecting solerope to spring	7
Stopper	W Scot	Pots	BUTTON	
Stopper link	Fleetwood Hull	Trawl	STOPPER	
Stopping(s)	S Eng	Trawl	SETTING(S), seizing fishing line to groundrope	
STORE POT			Large pot with door but no funnel or eye, in which crabs or lobsters are held live. *Syn* corb (Cornwall); keep pot (SW Eng)	
STORM LINES		Trawl	Strengthening ropes running diagonally across the square. *Syn* breastropes (Grimsby); lacing lines (NE Eng); square lines (NE Eng)	
Stotter	W Scot SW Scot	Trawl	BUMPER BOBBIN	
STOW NET			Conical net held open by one or more horizontal beams below an anchored boat for sprats, whitebait	
Stowing rope	SW Scot	Trawl	GROUNDROPE of a beam trawl	
Straight net	Dorset		SET GILL NET	
Straight selvedge	Literature	Net making	DOUBLE SELVEDGE	
STRAND		Rope	One of the main components of rope, each of which is made by twisting together filaments of wire or fibre	
Strap	1 S Eng 2 N Ire	Trawl Drift net	STIFFENER BUOY ROPE	
Strapping	S Eng	Trawl	STIFFENER	
Stray line	S Eng	Pots	BACKROPE	
Streamer	Cornwall	Set net	Small inflated buoy on separate line fixed to dan or dan line to aid recovery	
Strengthener	1 NE Eng 2 Grimsby	Trawl Trawl	STIFFENER DIVISIONAL BAR	
Stretch	Moray Firth	Longline	LONGLINE	
Stretchers	SW Scot	Pots	CROSS STICKS	
Striking up line	S Eng	Gill net	STAPLING	
STRING		1 Longline	One snood-bearing section of longline. *Syn* basket (E Eng); bath (Lowestoft); line (Eng); piece (NE Eng); tub (E Eng); unit (E Eng)	31
		2 Pots	A number of pots joined by branch lines to the backrope. *Syn* fleet (NE Eng); tear, tier	30

Term	Locality	Gear Type	Definition	Fig
			(Cornwall); trot (SW Eng); trotted pots (SW Eng); unit (E Eng)	
STRONGBACK	Fleetwood	Trawl	Strengthening rope from toe end of top wing to fishing line in the bunt	
Strongback	Hull Grimsby	Trawl	BOLT ROPE	
Strood	SE Scot	Longline	SNOOD	
STROP			Short length of wire or fibre rope usually with an eye splice at each end	
Strop	1 Lowestoft SE Scot	Drift net	BUOY ROPE	
	2 Granton SE Scot W Scot	Pots	POT STROP	
	3 Cornwall	Longline Handline	SNOOD	
	4 Grimsby	Trawl	Strengthening rope encircling the codend of a midwater trawl	
STRUTS	SE Eng	Dredge	Wire loops or clips linking ringles together to form chain belly	
Strutting	SE Eng	Pots	CROSS STICKS	
Stud	Devon	Pots	RIB	
SUBERKRUB OTTER BOARD		Trawl	All-steel cambered midwater otter board with vertical aspect greater than its horizontal aspect (named after its inventor)	20
Suckrope	Moray Firth	Drift net	SOLEROPE	
SUNK GILL NET			Set gill net rigged so that it fishes well below the surface. *Syn* anchor net, bottom gill net (SE Scot)	5
SUNK NET		Drift net Set gill net	1 Net fished below the messenger 2 Loosely, any gill net fished with headline below the surface	
SUPPORT ROPES		Trawl	Longitudinal strengthening ropes fixed outside the codend of a midwater trawl and joined to each other and to bolt ropes by strops	
SURFACE GILL NET			Set gill net rigged to fish at or near the surface. *Syn* anchor surface net (Moray Firth)	5
Swag bag	Grimsby	Trawl	CODEND, extra long. See Money box	
Swallow	Sussex	Trawl	FUNNEL and EXTENSION PIECES joined	
Swallow piece	SW Scot	Trawl	EXTENSION PIECE, of beam trawl	
Swallow tail	NE Scot	Trawl	FISH TAIL	
SWEEP		1 Danish seine	One of two ropes usually of combination rope connecting dan leno to the head line or fishing line. *Syn* bridle (Granton,	26

(cont)

Term	Locality	Gear Type	Definition	Fig
		2 Ring net	Grimsby); extension (W Scot); spreader (Moray Firth) The main towing rope connecting vessel to the net via the bridles. *Syn* canner (W Scot); end (W Scot); hook rope (W Scot); towing (SW Scot); towing line (SE Scot); towing rope (SW Scot); wee rope (W Scot)	7
Sweep	1 Dorset	Set gill net Trammel net	BRIDLE	
	2 SW Eng	Long line	TROT TOW	
	3 Scot	Trawl	BRIDLE	
Sweep net	SW Scot		BEACH SEINE	
Sweep pennant	Aberdeen	Trawl	PENNANT	
Sweep rope	Moray Firth	Purse seine	BUNT BRIDLE	
Sweep stopper	Aberdeen	Trawl	STOPPER	
SWING ROPE		Drift net	Rope connecting vessel to first pilchard fly net or salvagee	5
SWIVEL		General	Two links joined end to end by a pivot, thus allowing one link to rotate without the other	24
Swivel and flat link	Hull	Trawl	SWIVEL TOWING CHAIN WITH RECESSED LINK	
Swivel and O-link	Aberdeen	Trawl	SWIVEL TOWING CHAIN WITH RECESSED LINK	
Swivel and split link	Common	Danish seine	SWIVEL WITH CLIP LINK	
Swivel eye and link	NW Scot	Trawl	SWIVEL TOWING CHAIN	
Swivel link	1 NE Eng	Trawl	SWIVEL TOWING CHAIN WITH RECESSED LINK	
	2 SE Scot Moray Firth	Trawl	SWIVEL TOWING CHAIN	
Swivel piece	SE Scot	Trawl	SWIVEL TOWING CHAIN	
Swivel shoe	Aberdeen	Trawl	SWIVEL TOW SHACKLE	
SWIVEL TOW SHACKLE		Trawl	Heavy shackle fixed to the aft edge of an otter board and with a swivel eye for connection to legs, frame rope or tickler chain. *Syn* swivel shoe (Aberdeen); toe eye (Granton); toe shackle (common); tow shackle (Moray Firth)	21
SWIVEL TOWING CHAIN		Trawl	Two or three links of chain with swivel shackled between the brackets and warp. *Syn* board swivel link (Granton); board towing chain (Aberdeen); door chain (Aberdeen); door chain and swivel (Aberdeen, Moray Firth); door swivel chain (Aberdeen); swivel eye and link	

Term	Locality	Gear Type	Definition	Fig
			(NW Scot); swivel link (SE Scot, Moray Firth); swivel piece (SE Scot); towing link (SW Scot); towing swivel (Scot); trawl board chain and swivel (Aberdeen)	
SWIVEL TOWING CHAIN WITH RECESSED LINK		Trawl	A towing chain comprising swivel, link and recessed link fixed to end of warp to engage in a G-hook shackled to the brackets of an otter board. *Syn* G-hook swivel and link (Aberdeen); G-link and swivel (Moray Firth, NW Scot); swivel and flat link (Hull); swivel and O-link (Aberdeen); swivel link (NE Eng); warp end swivel (Hull, Grimsby)	21
SWIVEL WITH CLIP LINK			Combined swivel and clip link. *Syn* seine net swivel (Aberdeen); swivel and split link (common) swivel with link (Aberdeen, SE Scot); swivel with seine link (Scot)	
Swivel with link	Aberdeen SE Scot	Danish seine	SWIVEL WITH CLIP LINK	
Swivel with seine link	Scot	Danish seine	SWIVEL WITH CLIP LINK	
Sword	W Scot SW Scot	Dredge	TEETH BAR	
T-DIRECTION		Net making	The direction in netting parallel to the general course of the nettng yarn. Ref BS 4440:1974	1
TACHINGS	Cornwall	Drift net	Short lashings connecting the gables of adjacent pilchard nets as scunnings. *Syn* squinches (Cornwall)	
Tail	1 W Scot	Pots	POT STROP	
	2 Moray Firth W Scot	Trawl	EXTENSION PIECE	
	3 Granton	Trawl	Codend plus extension piece	
Tail end	Granton	Danish seine Trawl	CODEND	
Tail piece	Moray Firth	Trawl	EXTENSION PIECE	
Take ups	Grimsby	Net making	CLEAN MESHES	
TANGLE NET			Large-meshed loosely hung single panel set net which catches fish by entangling them	
Taper	Shetland	Trawl	EXTENSION PIECE	
Targle	Cornwall	Drift net	BUOY ROPE	
Tayse buoy	Cornwall	Drift net	POLE END BUOY	
Tayse end	Cornwall	Drift net	POLE END	
Teal bobbin	Common	Trawl	CYLINDRICAL RUBBER BOBBIN (Teal is a trade name)	

Term	Locality	Gear Type	Definition	Fig
Tear	Cornwall	1 Pots 2 Drift net 3 Longline	STRING FLEET LONGLINE	
Teeth	Moray Firth SW Scot	Dredge	TEETH BAR	
Teeth and blade	Granton	Dredge	TEETH BAR	
TEETH BAR		Dredge	A rake-like steel bar, with teeth, bolted across the bottom of the frame and to which the belly is attached. *Syn* sword (W Scot, SW Scot); teeth (Moray Firth, SW Scot); teeth and blade (Granton)	28
TELLING	W Scot	Longline	Marker buoy used to mark where the gear changes course	31
Thief net	Lowestoft	Drift net	SCUM NET	
THIMBLE		General	Grooved metal or plastic fitting to protect the eye of an eye splice, heart-shaped for light duty, egg-shaped for heavy duty	25
Third door	W Scot	Pound net	SMALL DOOR	
Third entrance	SW Scot	Pound net	SMALL DOOR	
Thread	1 W Scot 2 SW Scot	Danish seine Drift net	SEINE ROPE FLEET, usually anchored	
Three bridle butterfly	Aberdeen	Trawl	DAN LENO TRIANGLE	
THREE BRIDLE TRAWL			Trawl with bridle system incorporating three ropes each side connected to net headline, lestridge and groundrope, and usually without dan leno	17
Three legger	S Eng	Net making	HALFER	
Through bar	Fleetwood	Trawl	DIVISIONAL BAR	
THRUMS		Trawl	Strands of rope looped round bars of belly and codend meshes for protection against chafing. See also CHAFER	
TICKLER CHAIN		Trawl	Chain towed in front of groundrope to disturb flatfish	23
Tie	Granton	Trawl	GROMMET	
Tie bar	Hull Grimsby	Trawl	DIVISIONAL BAR	
Tier	Cornwall	1 Gill net 2 Longline 3 Pots 4 Trawl	FLEET LONGLINE STRING GROMMET	
Tin hat float	Grimsby	Trawl	TRAWL PLANE	
Tip	W Scot	Longline	TIPPING	
TIPPING		Longline	Thin single line joining hook to snood. *Syn* cotton (Moray Firth, SW Scot); graith	31

Term	Locality	Gear Type	Definition	Fig
			(Moray Firth); hemp (SE Scot); imp (SE Scot); tip (W Scot)	
TISSOT		Drift net	Rope or chain made fast to messenger when lying to nets, which takes the strain and saves chafing of messenger against the vessel	5
TOE		Trawl	Net section at the leading end of top or lower wings, usually of stronger or double twine. *Syn* point (Hull); toe end (NE Scot); toe piece (Aberdeen, NE Scot); wing point (Hull)	9 12
Toe bobbin	Moray Firth	Trawl	BUMPER BOBBIN	
Toe chain	Lowestoft	Trawl	BOTTOM LEG, of chain	
Toe end	NE Scot	Trawl	TOE	
Toe end becket	Granton	Danish seine	WING LINE	
TOE END WEIGHT		Trawl	Heavy weight, usually of iron or chain, at toe end of lower wing or on lower leg near wing of midwater trawl. *Syn* weight (Lowestoft)	18
Toe eye	Lowestoft	Trawl	SWIVEL TOW SHACKLE	
Toe leg	Hull Grimsby Fleetwood	Trawl	BOTTOM LEG	
Toe piece	Aberdeen NE Scot	Trawl	TOE or FISHTAIL	
Toe rope	Moray Firth	Danish seine	WING LINE	
Toe shackle	Common	Trawl	SWIVEL TOW SHACKLE	
Toe wire	Fleetwood	Trawl	BOTTOM LEG	
TOGGLE		Trawl	Short chain with a larger link at one end enclosing a heavier iron ring. Used to attach a rubber groundrope to the fishing line. *Syn* chain link (W Scot); cringle (Aberdeen, Granton); double link (W Scot); drop (NW Scot); drop chain (W Scot); footrope link (Aberdeen, W Scot); Italian towing chain (SE Scot); ring (W Scot); two chain link (Moray Firth)	15
Top and bottom gear	S Eng	Trawl	With otter boards connected directly to the net	
Top and bottom bag laced	Moray Firth	Trawl	BELLY AND BATINGS joined	
Top bag	1 Scot 2 Scot	Ring net Trawl	TOP BUNT BATINGS	
Top belly	1 Grimsby 2 Moray Firth	Trawl Trawl	TOP PANEL BATINGS	

Term	Locality	Gear Type	Definition	Fig
Top blade of wing	S Eng	Trawl	TOP WING	
Top bridle	1 Granton 2 SE Scot W Scot	Danish seine Trawl	TOP SWEEP TOP LEG	
TOP BUNT		Purse seine Ring net	Upper net section of the bunt. *Syn* basketing sling (W Scot); dryer (N Ire); top bag (Scot)	6
Top cheam	Cornwall	Pots	CANE HOOP, the top one	
Top cord	W Scot	Drift net	HEAD CORD	
Top corner	Moray Firth	Trawl	TOP QUARTER	
Top crouping	W Scot	Trawl	TOP QUARTER	
Top doubling	W Scot	Ring net	DOUBLE HEADLINE	
Top extension	W Scot	Danish seine	TOP SWEEP	
TOP EXTEN-SION PIECE		Trawl	EXTENSION PIECE, of top panel	9 10 11 23
Top fish tail	NE Scot NW Scot	Trawl	TOP TOE	
Top frame	Grimsby	Trawl	COPE BAR	
Top gathering	W Scot	Trawl	TOP QUARTER	
Top goring	W Scot	Trawl	TOP QUARTER or TOP GUSSET	
Top half	Aberdeen Lowestoft	Trawl	Headline, square, top wings and lower wings joined together	
TOP LEG		Trawl	Uppermost component wire of bridle system, often joining dan leno to headline. *Syn* headline leg (Eng); headline spreader (Granton); headline tow (S Eng); headline wire (Aberdeen); top bridle (SE Scot, W Scot); top spreader (Scot, Grimsby); top spreading wire (Scot); top sweep (Scot). See LEG	16 17
Top netting	Moray Firth	Dredge	BACK	
TOP PANEL		Trawl	Comprising all the net sections of the upper part of the trawl, viz top wings, square, batings, top extension piece. *Syn* top belly (Grimsby)	9 10 11 23
Top part	Hull	Trawl	A square and top wings joined	
TOP QUARTER		Trawl	See QUARTER. *Syn* headline quarter, (Grimsby); top corner (Moray Firth); top crouping (W Scot); top gathering (W Scot); top goring (W Scot)	9
Top reps	N Ire	Purse seine	HEADLINE	
Top sheet	1 NE Scot	Trawl	TOP WING	

Term	Locality	Gear Type	Definition	Fig
	2 Moray Firth SW Scot	Trawl	BATINGS	
Top spreader	1 Scot Grimsby 2 Moray Firth	Trawl Danish seine	TOP LEG TOP SWEEP	
Top spreading wire	1 Scot 2 Moray Firth W Scot	Trawl Danish seine	TOP LEG TOP SWEEP	
Top sticks	SE Scot	Pots	CROSS STICKS	
TOP SWEEP		Danish seine	Rope connected between end of headline and dan leno. *Syn* top bridle (Granton); top extension (W Scot); top spreader (Moray Firth); top spreading wire (Moray Firth, W Scot)	27
Top sweep	Scot	Trawl	TOP LEG	
Top swivel	Cornwall	Handline	BARREL BUCKLE SWIVEL	
TOP TOE		Trawl	Toe of top wing. *Syn* top/upper fish tail (Moray Firth, NE Scot); top toe end (Moray Firth); top toe piece (Aberdeen)	9 12
Top toe end	Moray Firth	Trawl	TOP TOE	
Top toe piece	Aberdeen	Trawl	TOP TOE	
TOP WING		Trawl	Net section extending forward from one side of the square and usually joined to the adjacent lower wing (two panel trawls) or adjacent side wing (four panel trawls). *Syn* peak (Granton); point (Granton); top blade of wing (S Eng); top sheet (NE Scot); upper wing (common)	9 10
TOP WING GUSSET		Trawl	Gusset at top quarter. *Syn* head gusset (Moray Firth). See GUSSET	12
TOPPING CHAIN		Trawl	First transverse chain of the stone mat fixed to beam at intervals	
TOPSIDE CHAFER		Trawl	Netting fixed outside the top section of codend to reduce chafing	
Torpedo	1 W Scot 2 Cornwall	Trawl Trawl	BUMPER BOBBIN See Retro-rocket	
Tow	1 W Scot 2 Wales	Pots Pots	BACKROPE POT LINE	
TOW BAR		Dredge	Steel bar or tube to which two or more dredges are attached. *Syn* dredge pole (SW Scot)	28
Tow leg	Hull Grimsby	Trawl	BOTTOM LEG	
Tow shackle	Moray Firth	Trawl	SWIVEL TOW SHACKLE	

Term	Locality	Gear Type	Definition	Fig
Towie	Moray Firth	Trawl	GROMMET	
Towing	SW Scot	Ring net	SWEEP	
TOWING CHAIN		Trawl	General term for the linkage system connecting the otter board to the warp and may include a few plain links, swivel, recessed link	21
Towing chains	Granton Moray Firth	Trawl	CHAIN BRACKET	
Towing line	SE Scot	Ring net	SWEEP	
Towing link	SW Scot	Trawl	TOWING CHAIN or SWIVEL TOWING CHAIN	
Towing rope	1 SW Scot 2 Moray Firth	Ring net Purse seine	SWEEP WEDGE ROPE	
Towing swivel	W Scot	Trawl	SWIVEL TOWING CHAIN	
Train	W Scot N Ire	1 Gill net Drift net 2 Longline	FLEET LONGLINE	
TRAMMEL NET			Set net having three net sections attached to common frame ropes. Usually the outer sections, the armouring, are of large meshes and the inner section, the lint, of small meshes	3
Trancher	Cornwall	Trammel net	ARMOURING	
TRAP		Pots	Hinged piece of wire across eye ring to form a one way trap	29
Trap	NE Eng	Trawl	FLAPPER	
Trap net	SE Scot		FYKE NET	
TRAWL			A fishing gear assembly incorporating a funnel-shaped net, ropes and hardware to hold open the mouth of the net when towed	8
Trawl board	Common		OTTER BOARD	
Trawl board bar	Aberdeen	Trawl	DIVISIONAL BAR	
Trawl board chain and swivel	Aberdeen	Trawl	SWIVEL TOWING CHAIN	
Trawl door	Common	Trawl	OTTER BOARD	
TRAWL HEAD		Trawl	Steel runner fixed to beam of a beam trawl, one at each end. *Syn* beam head (Devon); heel (Moray Firth); iron head (SW Scot); runner (Granton); shoe (Granton); sledge (SE Scot)	23
TRAWL PLANE		Trawl	Spherical metal headline float with an external flange for increased lift. *Syn* tin hat float (Grimsby); upthruster (Eng)	13

Term	Locality	Gear Type	Definition	Fig
Trawl warp	Common	Trawl	WARP	
Trawl warp marks	Common	Trawl	WARP MARKS	
Trawling wire	N Wales	Trawl	WARP	
Tret tow	E Eng	Longline	TROT TOW	
Triangle	1 Common 2 Moray Firth	Trawl Danish seine	BRACKET DAN LENO TRIANGLE	
TRIANGLE AND CHAIN BRACKETS		Trawl	On an otter board, a fore bracket and two chains in place of aft bracket	20
Triangle butterfly	Aberdeen	Trawl	DAN LENO TRIANGLE	
Triangle clip	Grimsby	Trawl	CLASP	
TRIPPER	NW Scot	Pound net	Line fixed to anchor of a bag net and floated at the surface. Used to break out the anchor before lifting it	
Tripping line	1 Grimsby 2 E Eng	Trawl Longline	LAZY DECKIE TROT TOW	
Trot	SW Eng	Pots	STRING	
Trot line	SW Eng	Longline	LONGLINE	
Trot rope	S Eng	Pots	POT LINE	
TROT TOW		Longline	Heavier line connecting end rope and dan line to the anchor. *Syn* spend tow (SW Scot); sweep (SW Eng); tret tow (E Eng); tripping line (E Eng); wash line (S Eng)	31
Trotted pots	SW Eng	Pots	STRING	
Trouser codend	Grimsby	Trawl	DOUBLE CODEND	
Tub	E Eng	Longline	STRING (one tub holds one string)	
Tunnel	S Eng	Pots	FUNNEL	
Twin beam gear	Devon	Trawl	DOUBLE BEAM TRAWL	
TWIN BRIDLE TRAWL			A trawl in which the net is connected to the otter boards by two long bridles each side. *Syn* double bridle trawl (Eng)	17
Twine net	Cornwall	Dredge	BACK	
Twisted bracket	S Scot	Trawl	SMALL BRACKET	
Two chain link	Moray Firth	Trawl	TOGGLE	
Tying down line	S Eng	Trawl	Rope, replacing top leg, which connects end of headline to forward end of chain bottom leg	
Underbridle	NE Eng	Pots	POT BRIDLE	

Term	Locality	Gear Type	Definition	Fig
Underblade	S Eng	Trawl	1 BELLY 2 LOWER WING	
Underheading	S Eng	Trawl	FISHING LINE	
Unit	E Eng	Longline	STRING	
Up and down bridle	Scot	Ring net	BRIDLE, the lower one	
Up and down line	S Eng	1 Trammel net 2 Trawl	GABLE WING LINE	
Upper blade	Sussex	Trawl	Top wings, square and batings joined together	
Upper fish tail	Moray Firth	Trawl	TOP TOE	
Upper heading	Sussex	Trawl	HEADLINE	
Upper wing	Common	Trawl	TOP WING	
Upthruster	Eng	Trawl	TRAWL PLANE	
V-BOARD		Trawl	All steel otter board with pronounced dihedral having only one bracket hinged horizontally. *Syn* steel door (Scot); V-door (common)	20
V-door	Common	Trawl	V-BOARD	
V-SHACKLE	Common	Trawl	V shaped shackle mainly used on dan leno spindle. *Syn* bobbin shackle (Aberdeen); Ross bobbin shackle, Ross shackle (Aberdeen)	24
V-piece	Devon	Trawl	SIDE PANEL	
V-wing	Grimsby	Trawl	FISH TAIL	
Valve	S Eng	Trawl	FLAPPER	
VD gear	Eng	Trawl	BRIDLE GEAR	
VD link	Hull	Trawl	BACKSTROP LINK	
VD ring	Cornwall	Trawl	KELLY'S EYE	
VD shackle	Aberdeen	Trawl	HEADLINE SHACKLE	
Vigneron Dahl gear	Eng	Trawl	BRIDLE GEAR	
Vinge trawl	Common	Trawl	WING TRAWL	
Wage	Moray Firth	Trawl	CODEND WEDGE	
Wall	Dorset	Trammel net	ARMOURING	
WALL OF NETTING			A panel of netting which is employed vertically	
Walling	S Eng	Trammel net	ARMOURING	
Warning mark	Hull	Trawl	SHOOTING MARK	
WARP		Trawl	Long flexible steel rope connecting vessel	8

Term	Locality	Gear Type	Definition	Fig
		Dredge	to the gear. *Syn* trawl warp (common); trawling wire (N Wales); wire (W Scot)	28
Warp	1 E Eng 2 Moray Firth W Scot 3 Lowestoft	Beach seine Danish seine Drift net	HAULING LINE SEINE ROPE MESSENGER	
Warp end swivel	Hull Grimsby	Trawl	SWIVEL TOWING CHAIN WITH RECESSED LINK	
WARP MARKS		Trawl Dredge	Marks, usually of fibre, inserted into lay of the warp and normally at 25 fathom intervals. *Syn* trawl warp marks (common)	
WARP SHACKLE		Trawl	Heavy D-shackle used mainly to connect the end of the trawl warp to the towing chain	21
Wash line	S Eng	Longline	TROT TOW	
Washer	1 Cornwall 2 Common	Pots Trawl	Float, often a plastic bottle ('bottle washer'), fixed to the buoy rope below the marker buoy in case the latter comes adrift DAN LENO WASHER	
Weather hole	S Eng	Pots	FUNNEL	
Webbing	1 SW Scot 2 Common	Drift net	LINT NETTING	
WEDGE	Scot	Purse seine	Triangular net section sometimes fitted to end of wing to lead bulk of net into power block at start of hauling	6
Wedge	N Ire	Trawl	CODEND WEDGE	
WEDGE ROPE		Purse seine	Hauling rope attached to wing end of net headline or to tip of wedge. *Syn* towing rope (Moray Firth)	7
Wedge shackle	Aberdeen		SHACKLE WITH FORELOCK PIN AND COTTER	
Wee rope	W Scot	Ring net	SWEEP	
Weight	1 W Scot 2 Moray Firth 3 Lowestoft	Longline Purse seine Trawl	SINKER LEAD TOE END WEIGHT	
Went	Dorset	Net making	ROW	
Wheel bobbin	Aberdeen Moray Firth	Trawl	CYLINDRICAL RUBBER BOBBIN	
Whiffing line	Cornwall	Handline	Very lightly weighted and baited trolling line	
Whipping twine			Any light twine suitable for whipping ends of rope to prevent fraying	
Wide net	W Scot	Ring net	WING	

Term	Locality	Gear Type	Definition	Fig
Wide wing	W Scot	Ring net	WING	
Windows	SW Scot	Trammel net	ARMOURING	
WING		1 Trawl Danish seine	Net section extending forward from one side of the main body of the net	9 10 11 12 23
		2 Ring net Beach seine	Net section at each end of the net. *Syn* end (W Scot); wide net (W Scot); wide wing (W Scot)	6
		3 Purse seine	Tapered net section(s) at the opposite end from the bunt	6
Wing bobbin	W Scot	Trawl	BUMPER BOBBIN	
Wing bunt bobbins	Moray Firth	Trawl	BUNT BOBBINS	
WING BUOY		Ring net	Inflated buoy attached to headline at junction of the wing and shoulder to mark that position	6
WING END		Trawl Danish seine	General term for leading end of wing	
WING END TICKLER		Trawl	Tickler chain the ends of which are attached to the forward ends of the groundrope. *Syn* wing tickler (Lowestoft)	
WING GUSSET		Trawl	Triangular net section at the top or lower quarters. *Syn* corner (Moray Firth); drops (Moray Firth)	12
WING LINE		Trawl Danish seine	Frame rope to which the leading edge meshes of the wing between the headline and fishing line are fastened, *eg* toe end meshes, fish tail meshes. *Syn* end line (NE Eng, N Ire); fishing line (Sussex); gable (Scot); gale (Danish seine, Moray Firth); gavel (Moray Firth); lug (W Scot); preventive (Granton); protector (Danish seine, Aberdeen); side heading (Sussex); side line (Devon); toe end becket (Danish seine, Granton); toe rope (Moray Firth); up and down line (S Eng); wing rope (Grimsby)	9 27
Wing point	Hull	Trawl	TOE	
Wing rope	Grimsby	Trawl	WING LINE	
WING RUBBERS		Trawl	Groundrope wing section clad with rubber discs and toggles. *Syn* rubber legs (Scot); rubbers (common)	14
Wing tickler	Lowestoft	Trawl	WING END TICKLER	
WING TRAWL			High opening two panel demersal trawl with pronounced fishtail. *Syn* vinge trawl (common)	

Term	Locality	Gear Type	Definition	Fig
Winkie	Scot	Purse seine Ring net	Dan with a battery operated flashing light for night fishing	
Wire	1 NW Scot 2 W Scot 3 Cornwall	Pound net Trawl Pots	CABLE WARP RIB	
Wire bellies	Cornwall	Dredge	CHAIN BELLY	
WIRE HEART		Trawl Danish seine	Groundrope assembled on wire as distinct from chain	
Wire pot	Cornwall	Pots	Cornish pot constructed of woven wire with no net cover	
Withy pot	Cornwall	Pots	CORNISH POT, made of withies	
Wolded groundrope	Fleetwood Lowestoft	Trawl	ROPE ROUNDED GROUNDROPE	
Wood bar	Cornwall	Dredge	BAR (one made of wood)	
Wooden bead	Lowestoft	Trawl	BOBBIN—cylindrical and made of wood	
WORMING		General	Following the lay of a wire rope with cord before it is parcelled and served	
WRAPPING CHAIN	Lowestoft	Trawl	Chain passed spirally around a rope or rope-rounded groundrope to improve ground contact. *Syn* chain wrapping (S Eng)	
Y-piece	Granton	Trawl	EXTENSION PIECE	
Yarn	E Scot	Drift net	LINT	
Yoke	Moray Firth	Trawl Danish seine	DAN LENO STICK	
Yoke hoop	Moray Firth	Danish seine	DAN LENO HOOP	
Yoke stick	Moray Firth	Danish seine	DAN LENO STICK	
Yook	NE Eng	Handline Longline	HOOK	

Figures

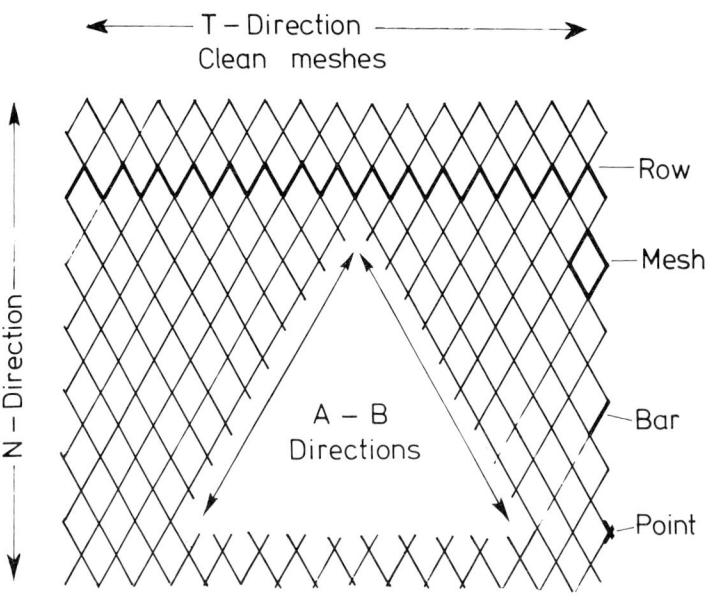

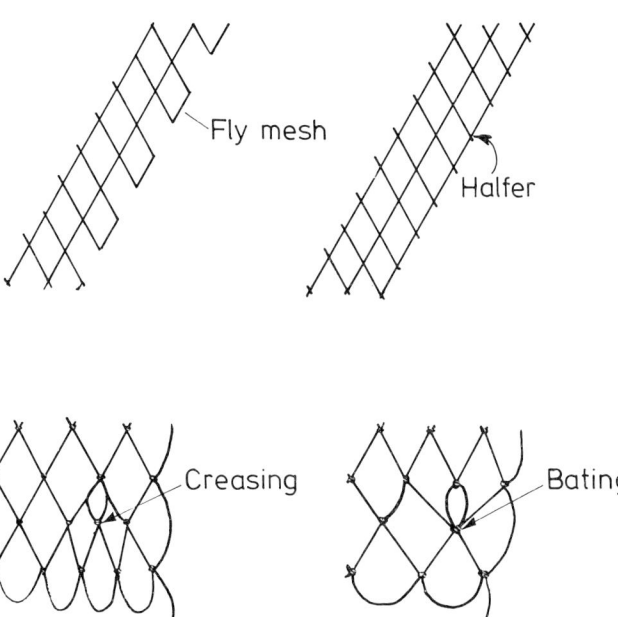

Fig 1 Netting—basic terms

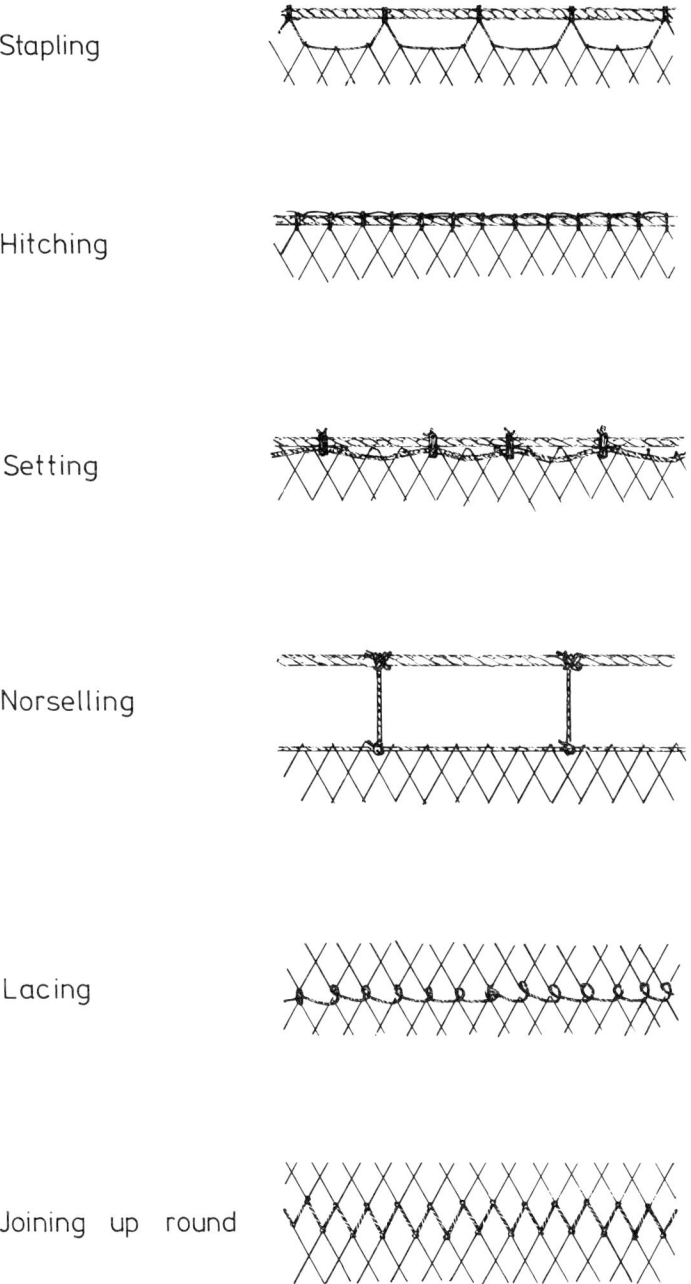

Stapling

Hitching

Setting

Norselling

Lacing

Joining up round

Fig 2 Joining and mounting of netting—basic terms

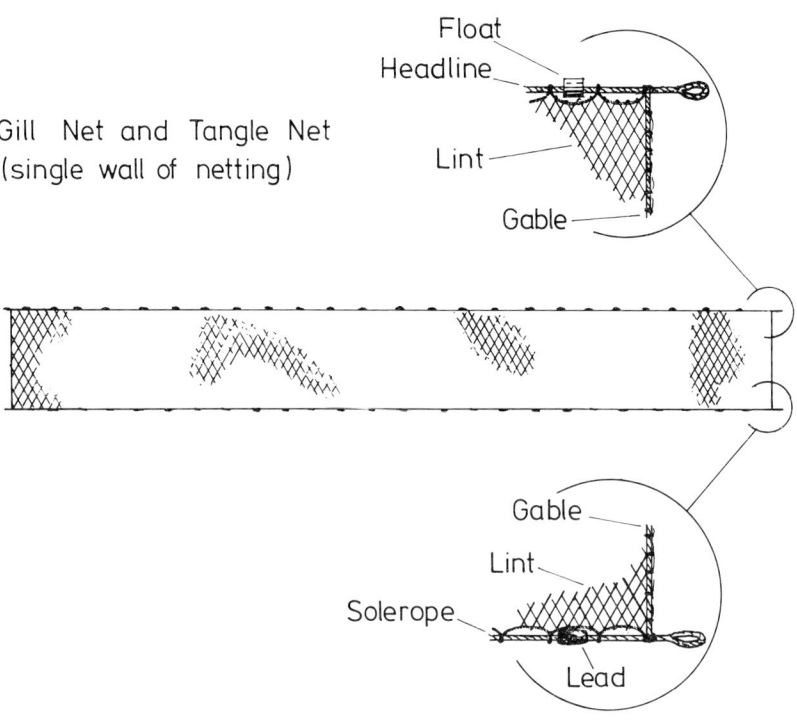

Gill Net and Tangle Net
(single wall of netting)

Float

Headline

Lint

Gable

Gable

Lint

Solerope

Lead

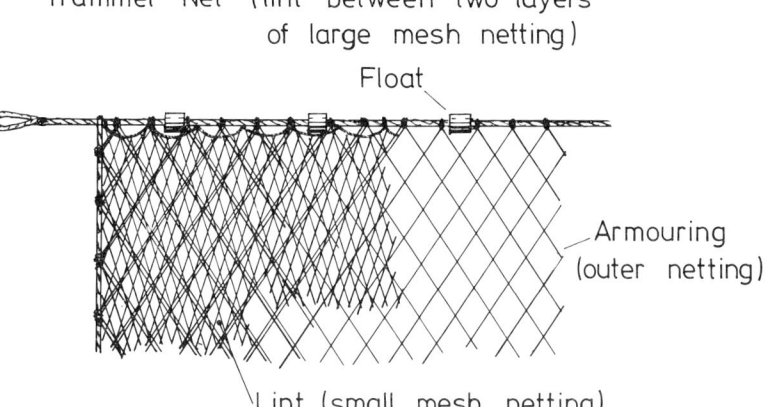

Trammel Net (lint between two layers
of large mesh netting)

Float

Armouring
(outer netting)

Lint (small mesh netting)

Fig 3   Gill, tangle and trammel nets

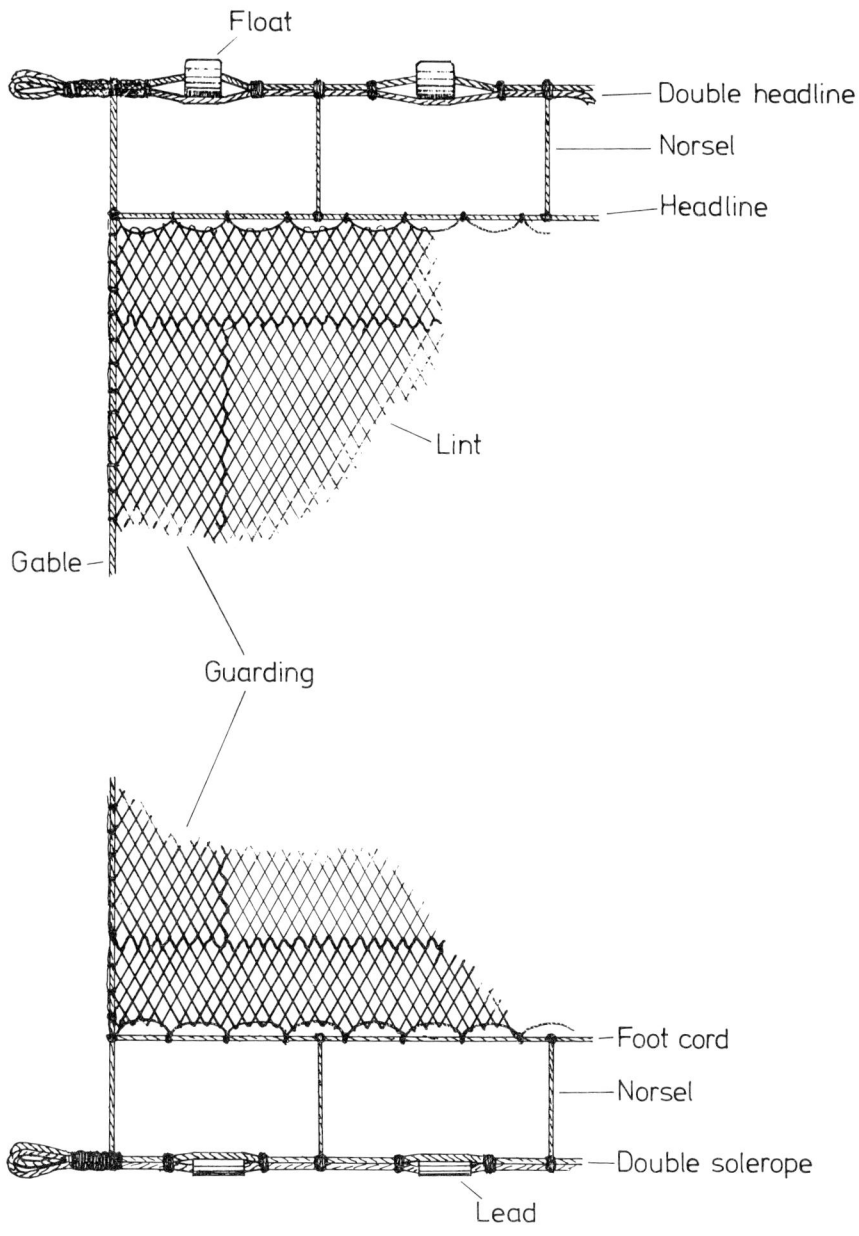

Float

Double headline

Norsel

Headline

Lint

Gable

Guarding

Foot cord

Norsel

Double solerope

Lead

Fig 4   Top and lower corners of a drift net

SET NET ON THE BOTTOM

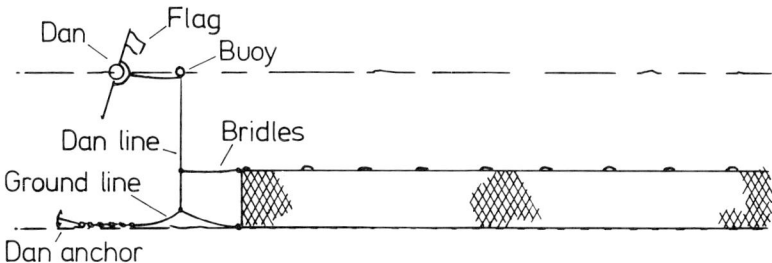

SET NET OFF THE BOTTOM

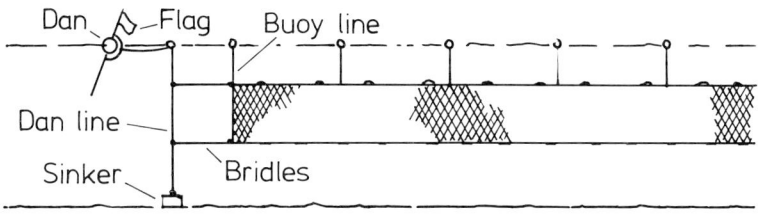

HERRING DRIFT NETS

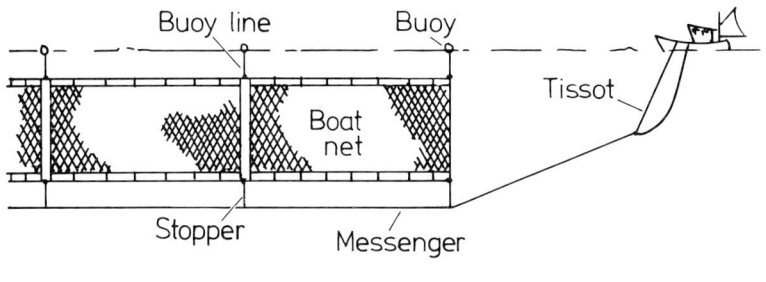

CORNISH PILCHARD FLY NETS

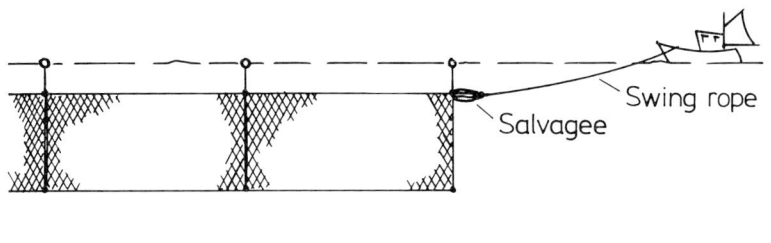

Fig 5   Various methods of fishing set gill nets and drift nets

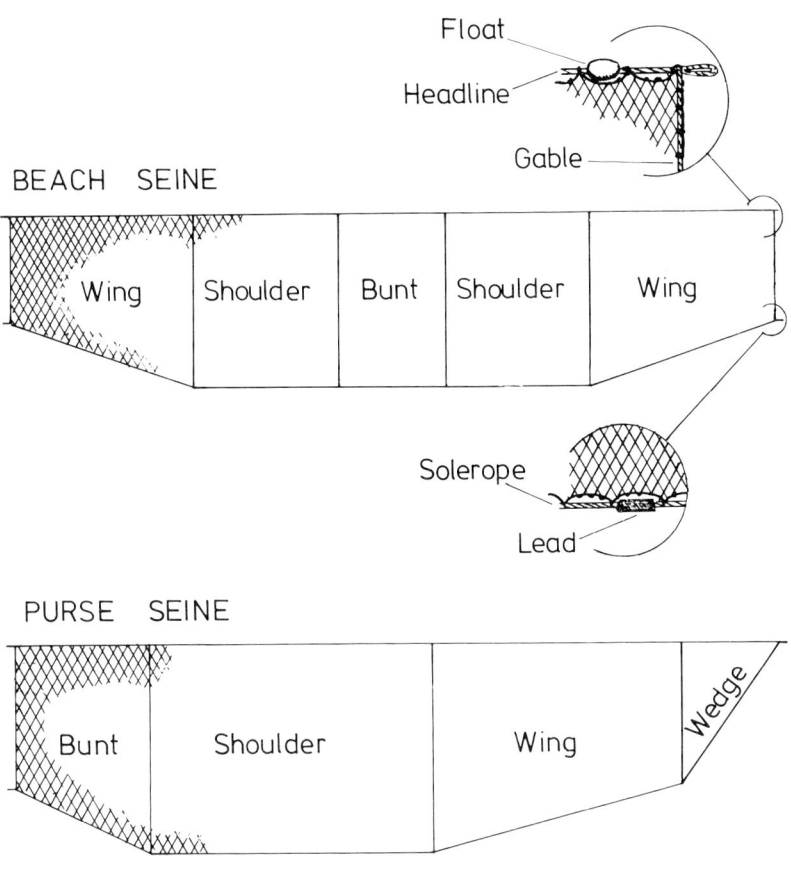

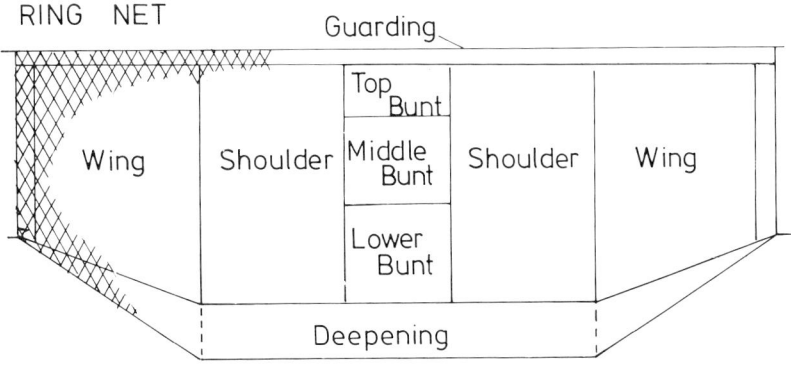

Fig 6 Principal features of a beach seine, purse seine and ring net. Headline, gables and solerope are also features of purse seines and ring nets

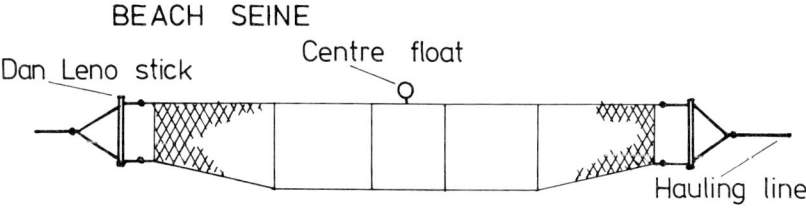

BEACH SEINE

Dan Leno stick

Centre float

Hauling line

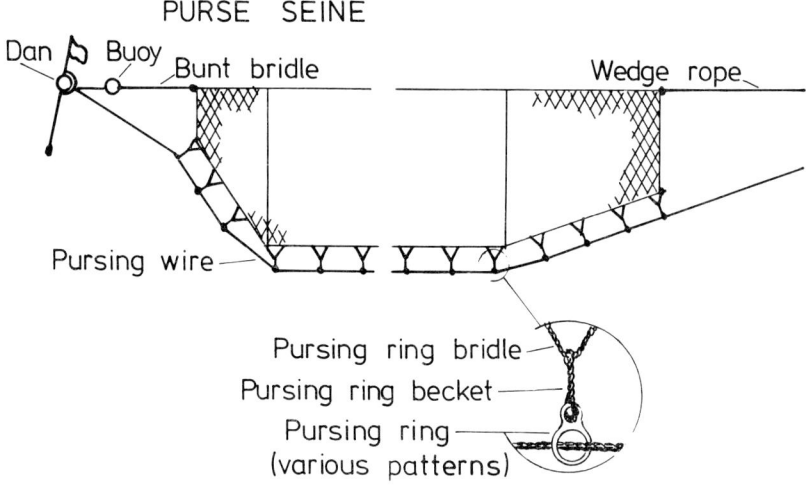

PURSE SEINE

Dan

Buoy

Bunt bridle

Wedge rope

Pursing wire

Pursing ring bridle

Pursing ring becket

Pursing ring
(various patterns)

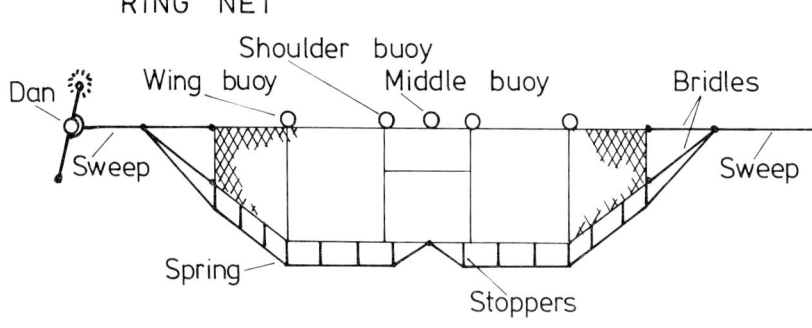

RING NET

Shoulder buoy

Wing buoy

Middle buoy

Bridles

Dan

Sweep

Sweep

Spring

Stoppers

Fig 7 Rigging items for beach seines, purse seines and ring nets

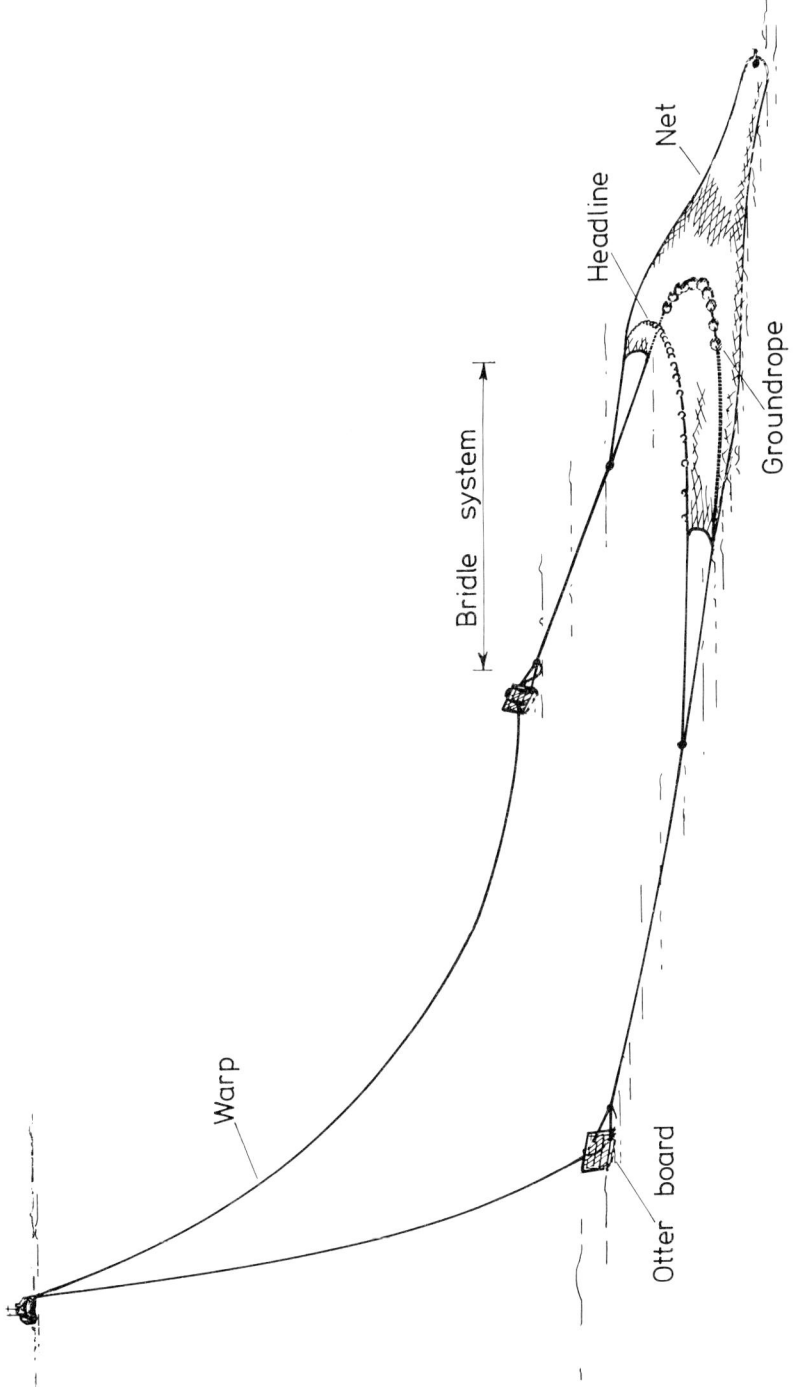

Fig 8 The principal features of a bottom otter trawl

Net

Headline

Groundrope

Bridle system

Warp

Otter board

89

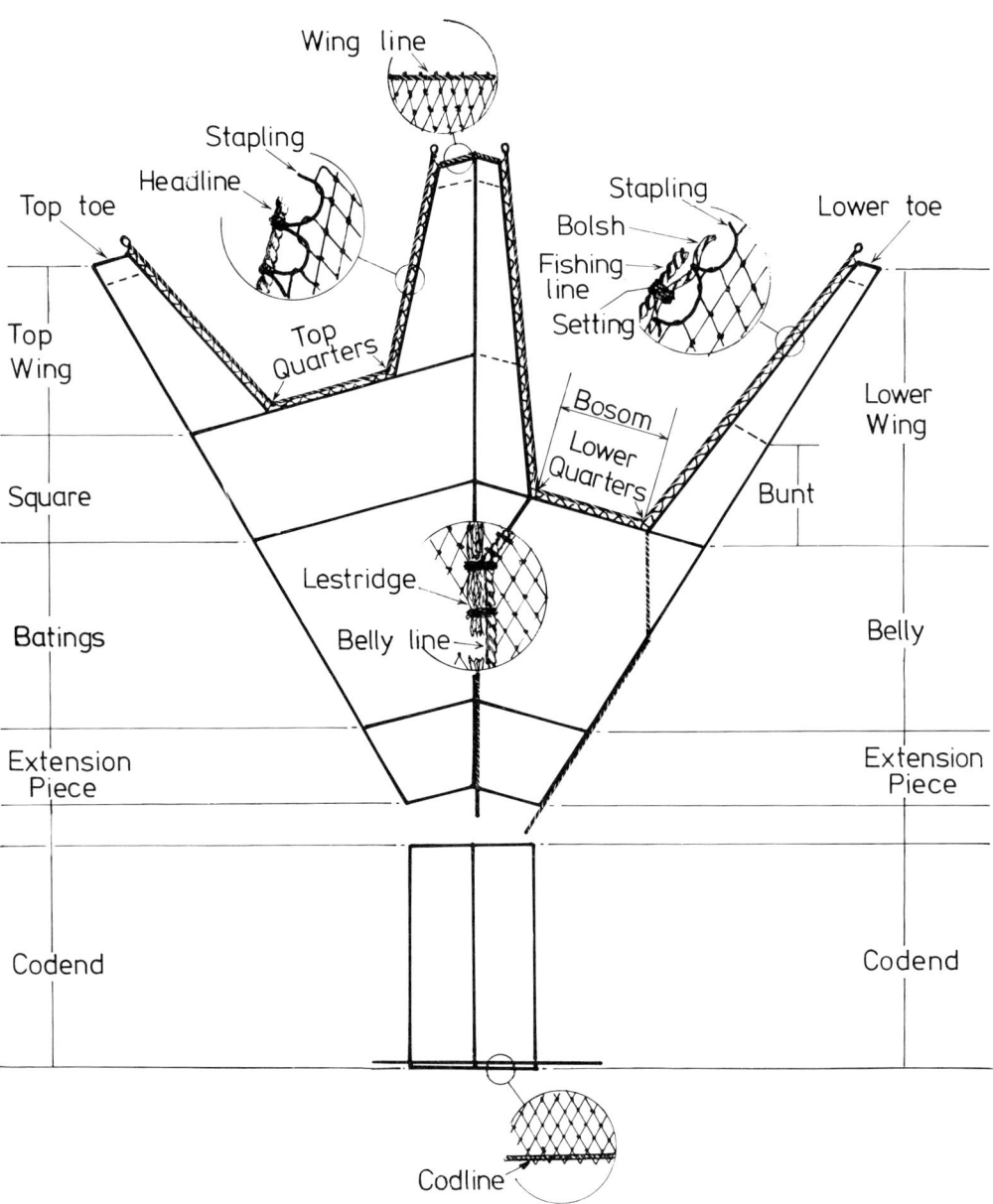

TOP PANEL

LOWER PANEL

Wing line

Stapling

Headline

Top toe

Top Wing

Square

Batings

Extension Piece

Codend

Stapling

Bolsh

Fishing line

Setting

Lower toe

Top Quarters

Bosom

Lower Quarters

Lestridge

Belly line

Lower Wing

Bunt

Belly

Extension Piece

Codend

Codline

Fig 9 The principal net sections and ropes of a two panel bottom trawl

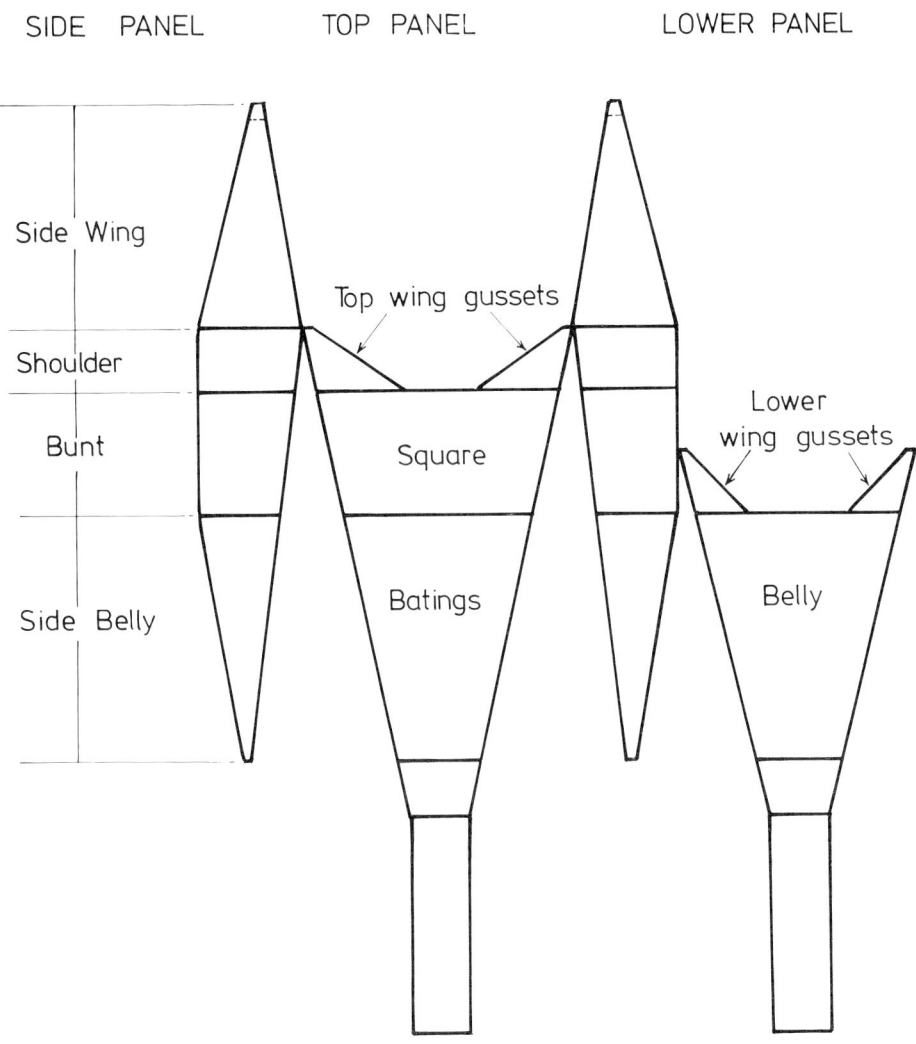

SIDE   PANEL   TOP PANEL   LOWER PANEL

Side Wing

Top wing gussets

Shoulder

Lower
wing gussets

Bunt

Square

Side Belly

Batings

Belly

Fig 10   The principal net sections of a four panel bottom trawl

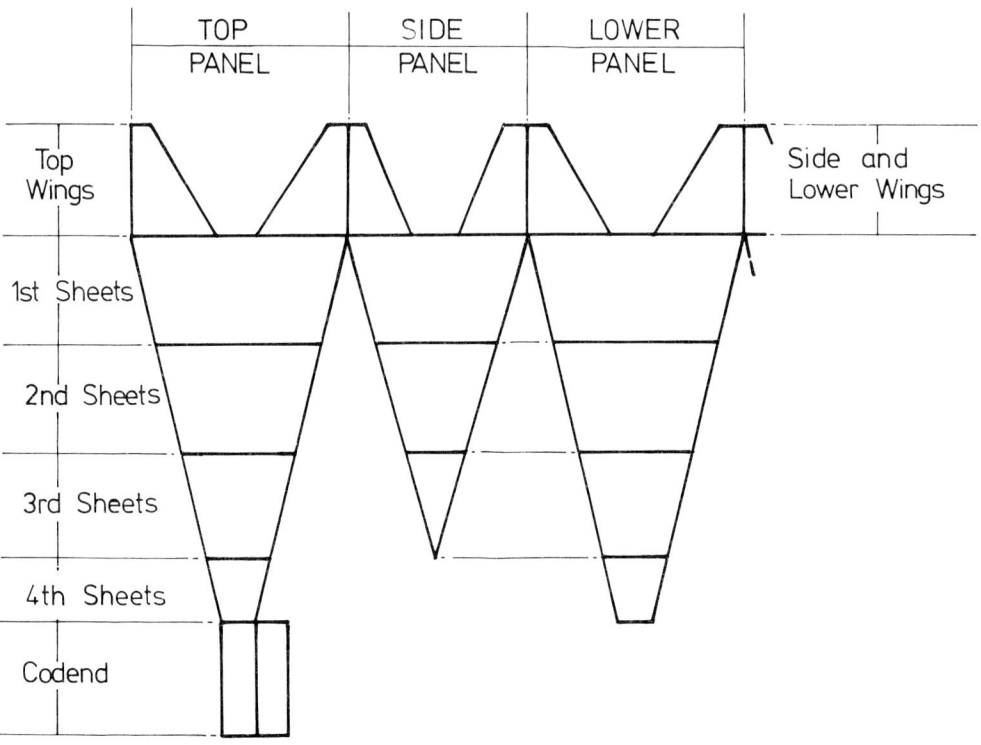

Fig 11   The principal net sections of a four panel midwater trawl. *Note*—the panels may contain fewer or more sheets than are illustrated

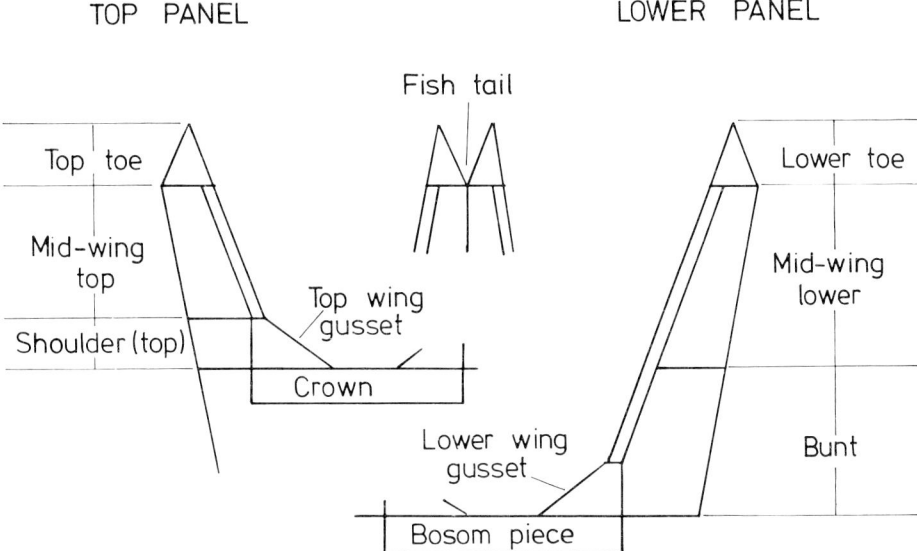

TOP PANEL

LOWER PANEL

Fish tail

Top toe

Mid-wing
top

Shoulder (top)

Top wing
gusset

Crown

Lower toe

Mid-wing
lower

Bunt

Lower wing
gusset

Bosom piece

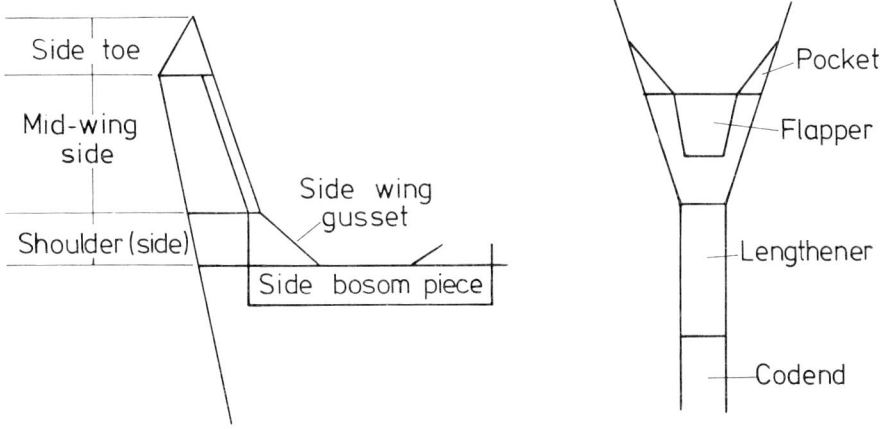

SIDE   PANEL

Side toe

Mid-wing
side

Shoulder (side)

Side wing
gusset

Side bosom piece

Pocket

Flapper

Lengthener

Codend

Fig 12   Net subsections that may feature in trawls and Danish seines

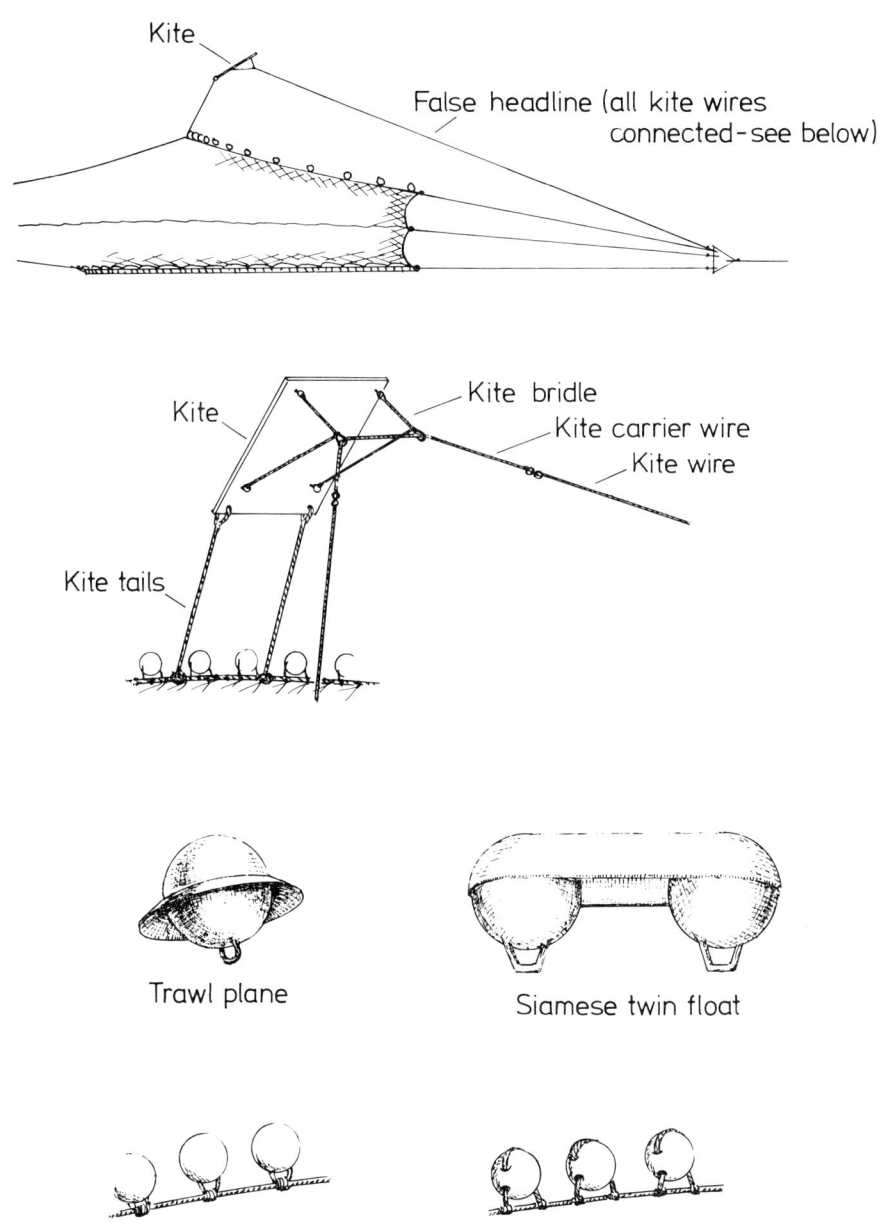

Kite

False headline (all kite wires connected – see below)

Kite

Kite bridle

Kite carrier wire

Kite wire

Kite tails

Trawl plane

Siamese twin float

Headline floats

Fig 13 Headline lifters

Spherical steel, rubber or plastic bobbins and spacers threaded on wires (bobbin wires)

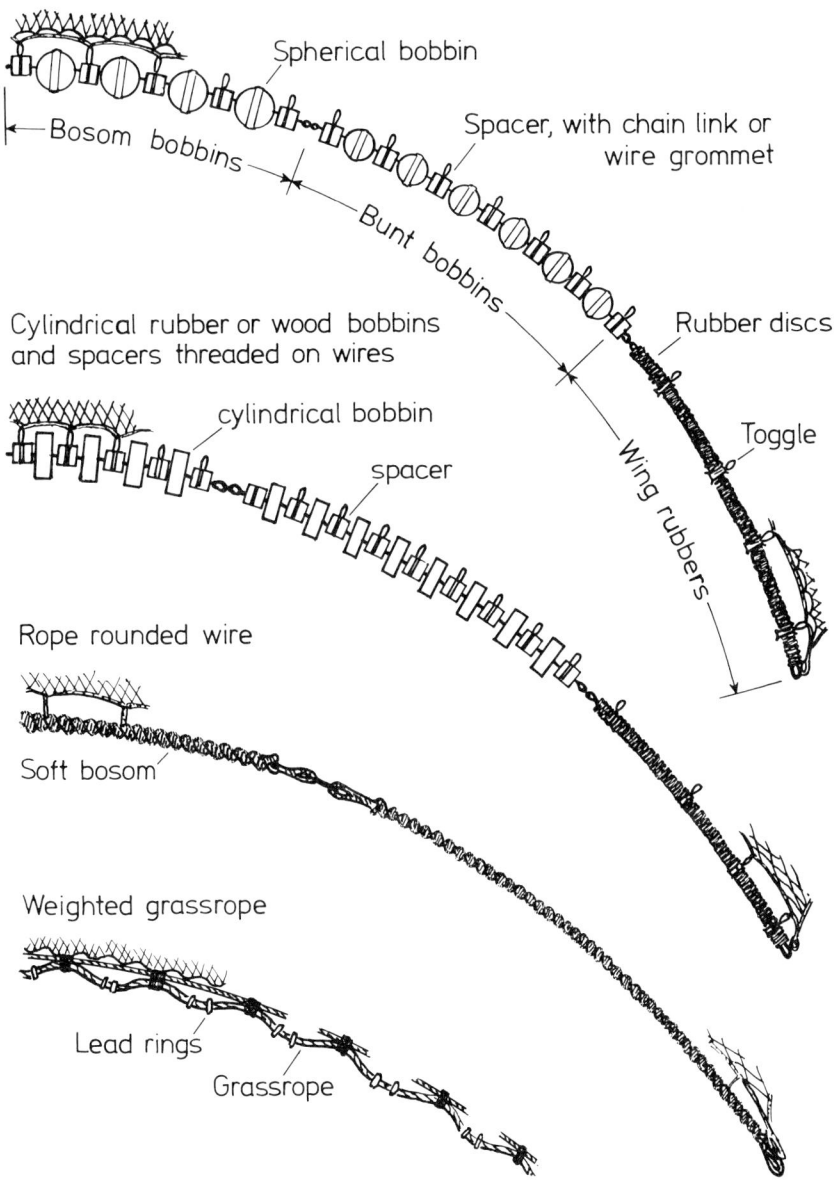

Spherical bobbin

Bosom bobbins

Bunt bobbins

Spacer, with chain link or wire grommet

Rubber discs

Wing rubbers

Toggle

Cylindrical rubber or wood bobbins and spacers threaded on wires

cylindrical bobbin

spacer

Rope rounded wire

Soft bosom

Weighted grassrope

Lead rings

Grassrope

Fig 14   Some types of groundropes

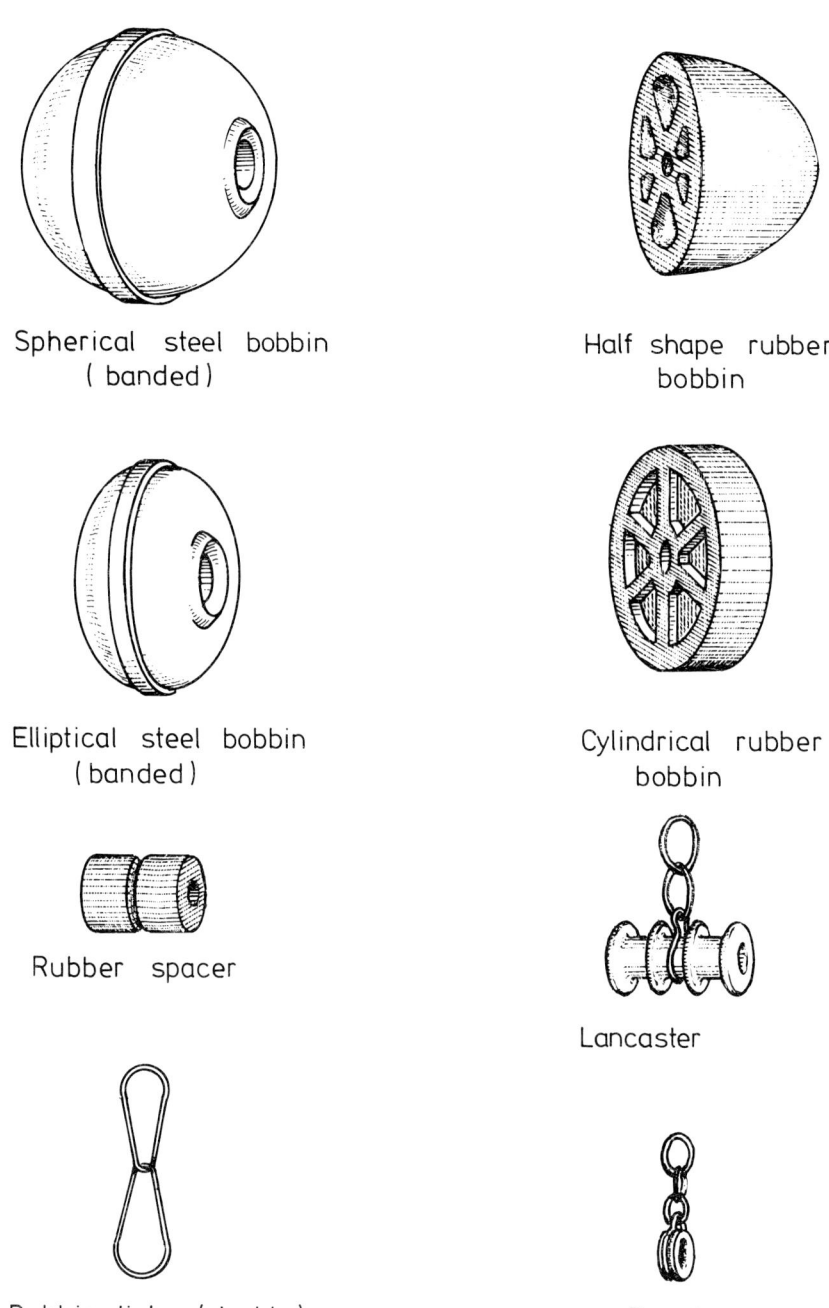

Spherical steel bobbin
(banded)

Half shape rubber
bobbin

Elliptical steel bobbin
(banded)

Cylindrical rubber
bobbin

Rubber spacer

Lancaster

Bobbin links (double)

Toggle

Fig 15   Some items used on groundropes

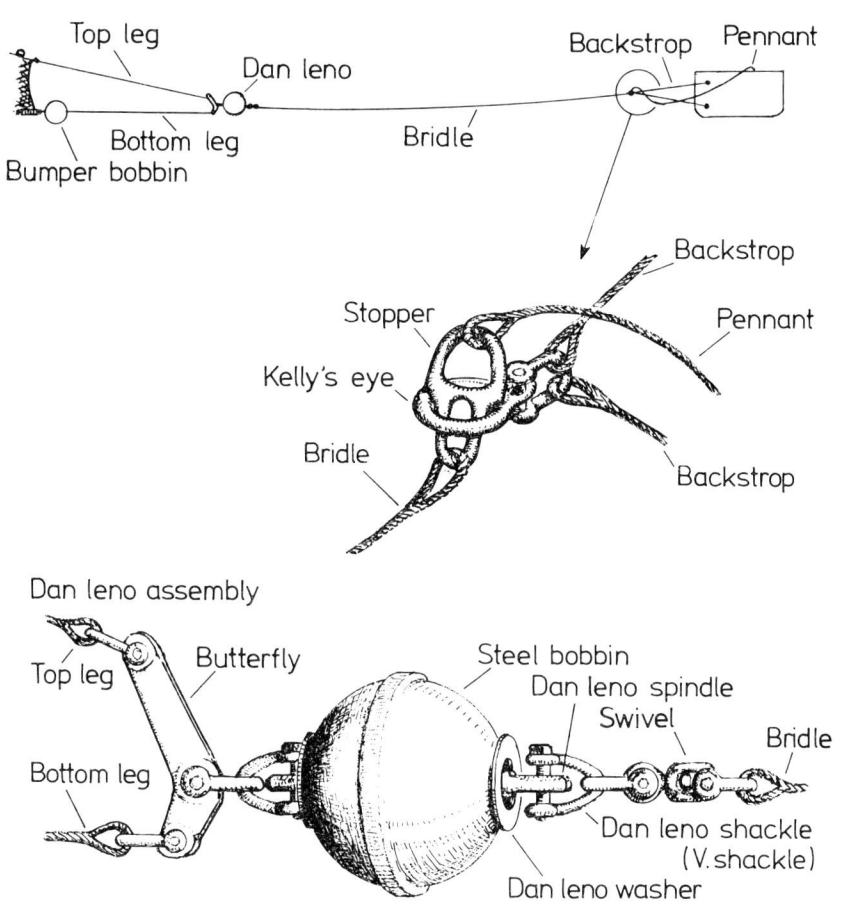

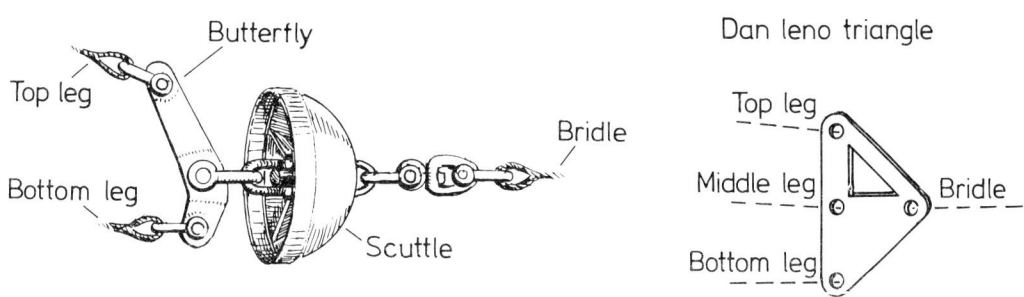

Fig 16   Principal features of a bridle system

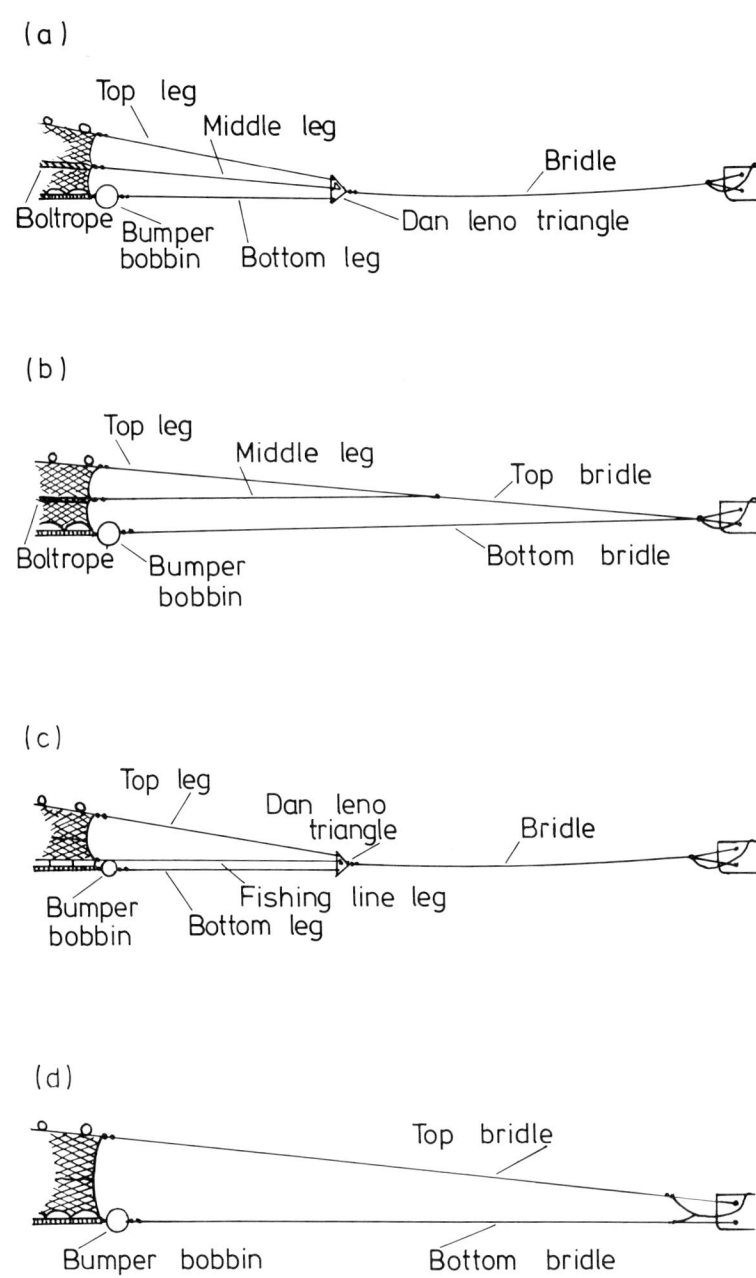

Fig 17 Various bridle systems for bottom trawling

(a)

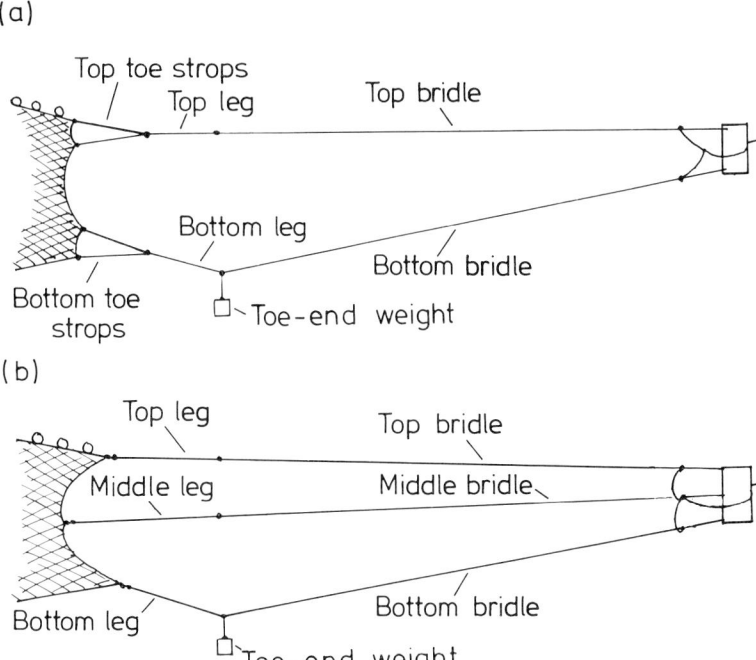

(b)

(c)

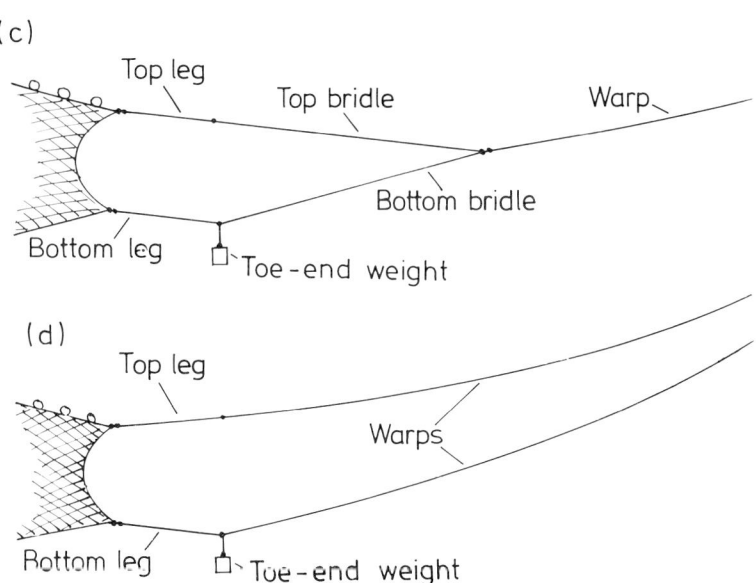

(d)

Fig 18 Some bridle systems for midwater trawling; for single boat—(a) twin bridles, (b) three bridles; for pair boats—(c) single warp, (d) twin warp

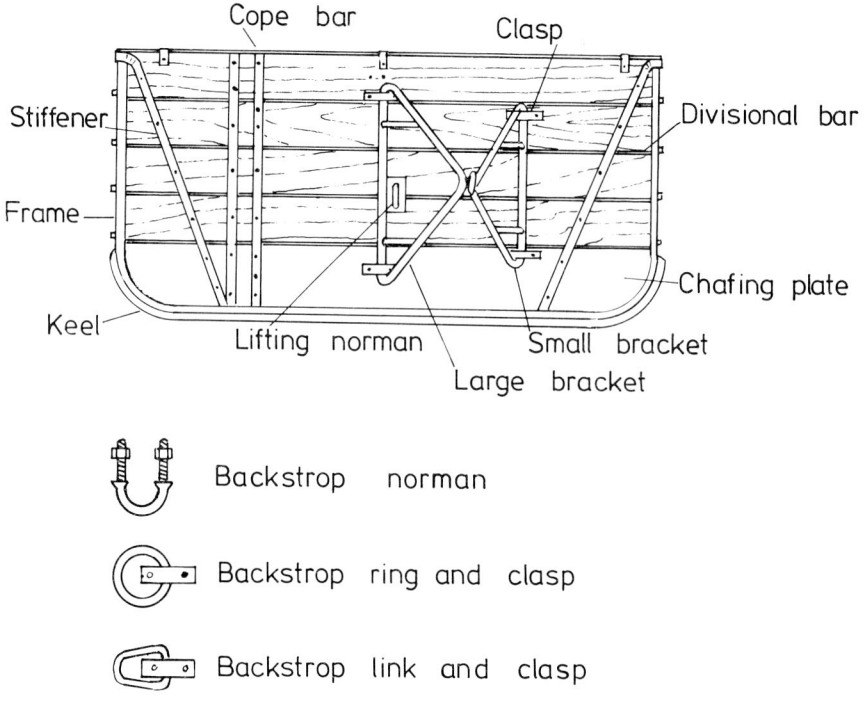

Backstrop norman

Backstrop ring and clasp

Backstrop link and clasp

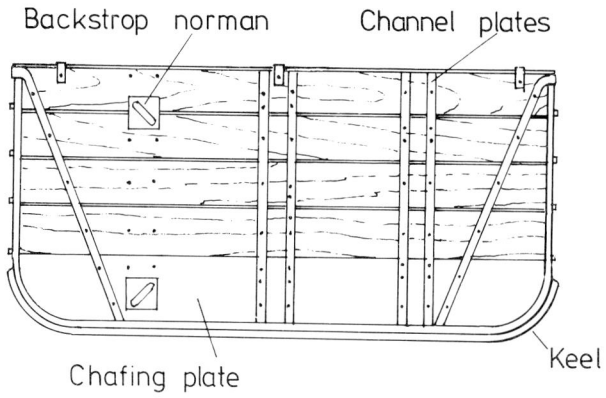

Fig 19 The main features of a flat otter board

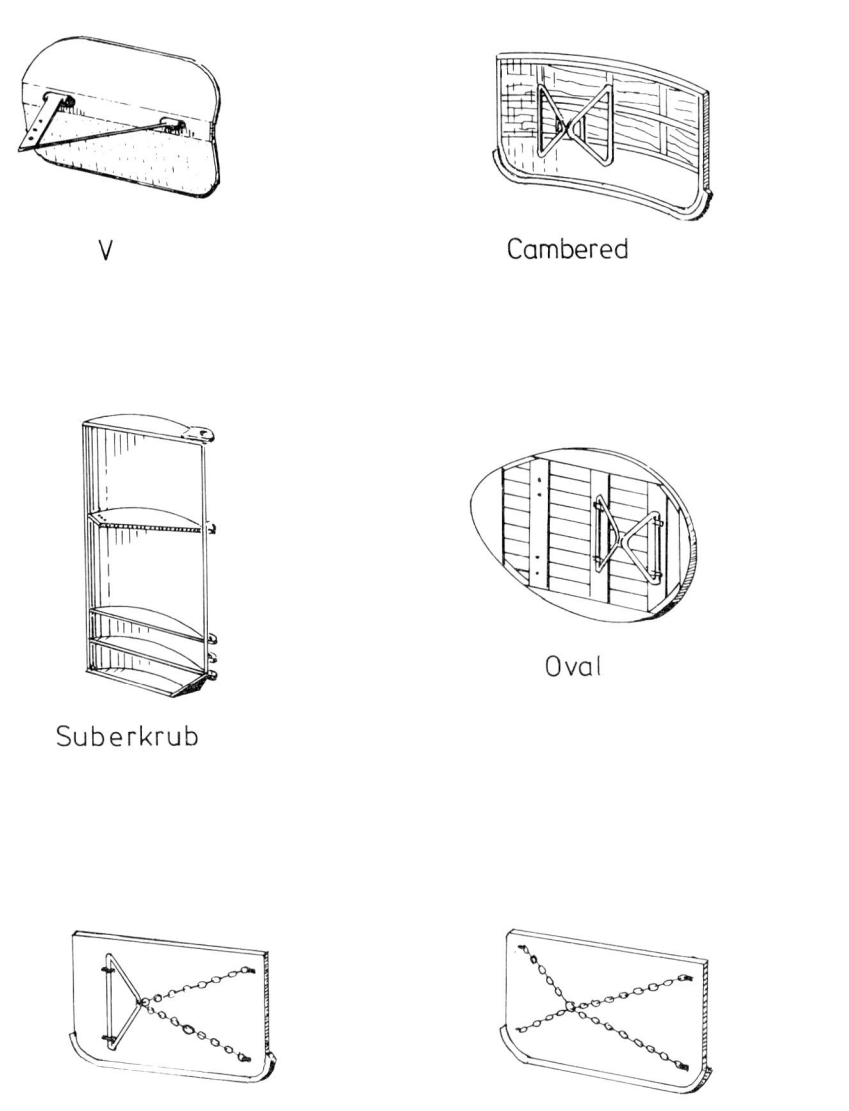

V

Cambered

Suberkrub

Oval

Flat – with triangle and
chain brackets

Flat – with chain brackets

Fig 20   Some types of otter boards

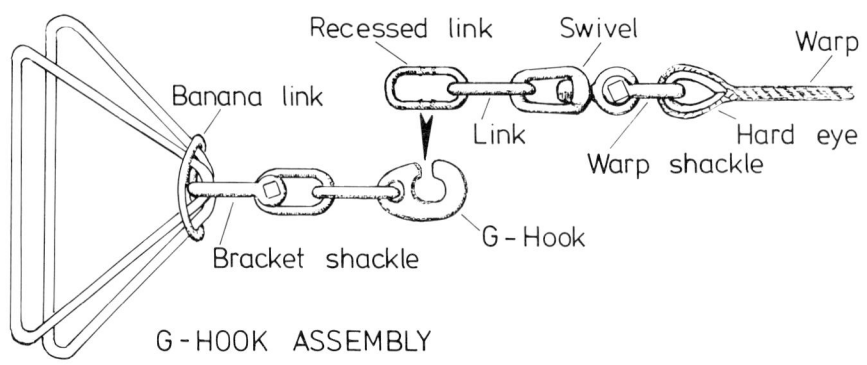

SWIVEL TOWING CHAIN WITH RECESSED LINK

Recessed link   Swivel   Warp

Banana link

Link

Hard eye

Warp shackle

Bracket shackle   G – Hook

G-HOOK ASSEMBLY

Swivel tow shackle

Backstrop equaliser

Fig 21   Otter board fittings for wire attachments

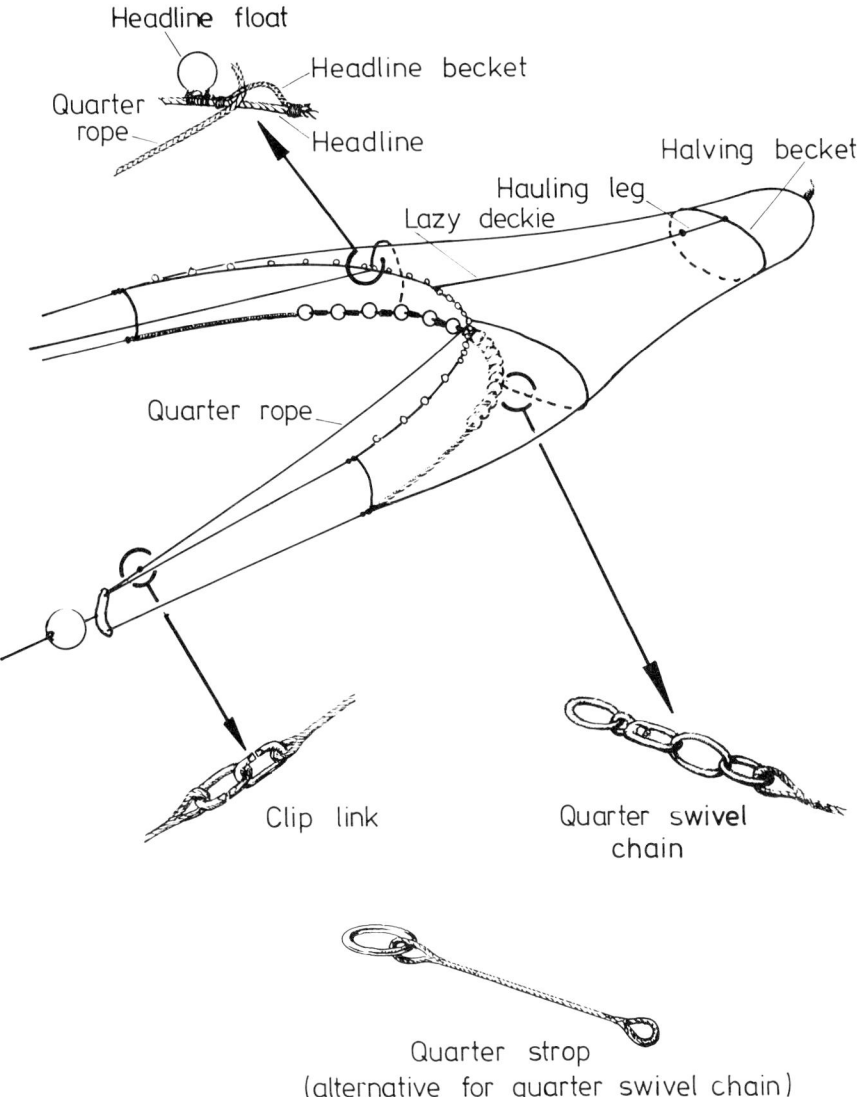

Headline float

Headline becket

Quarter rope

Headline

Halving becket

Hauling leg

Lazy deckie

Quarter rope

Clip link

Quarter swivel chain

Quarter strop
(alternative for quarter swivel chain)

Fig 22 Trawl handling gear

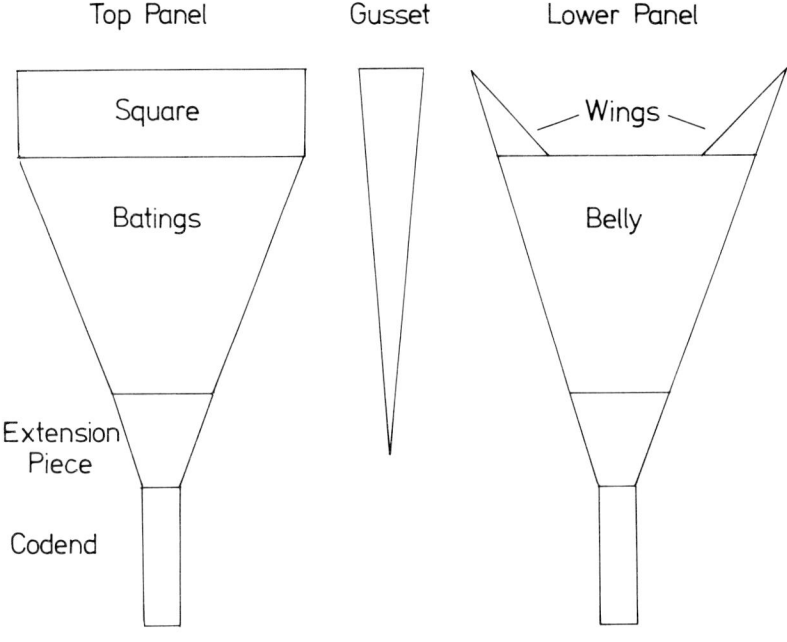

Top Panel

Gusset

Lower Panel

Square

Batings

Extension
Piece

Codend

Wings

Belly

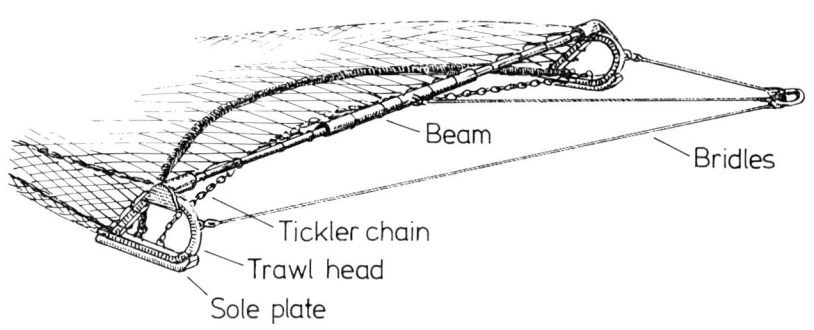

Beam

Bridles

Tickler chain

Trawl head

Sole plate

Fig 23 Beam trawl

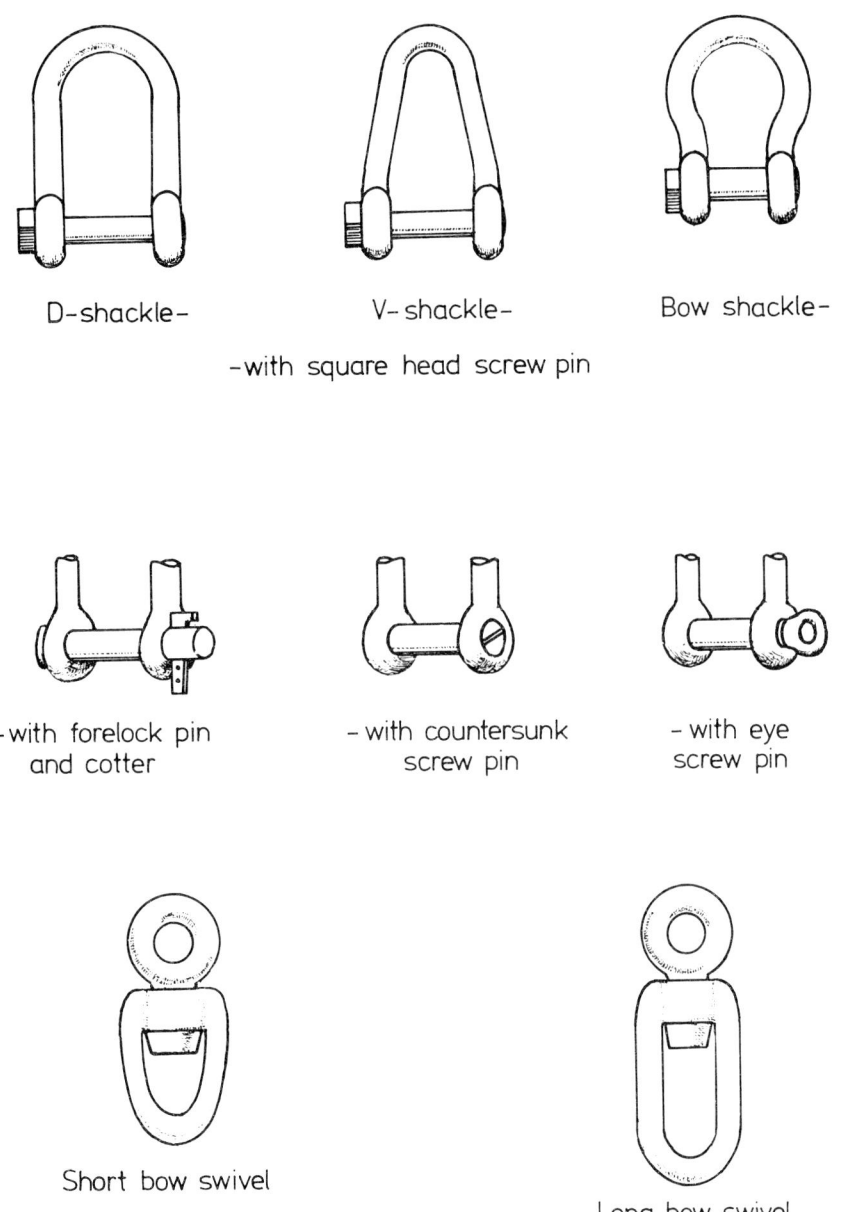

D-shackle-   V-shackle-   Bow shackle-

-with square head screw pin

-with forelock pin   - with countersunk   - with eye
and cotter   screw pin   screw pin

Short bow swivel

Long bow swivel

Fig 24   Shackles and swivels

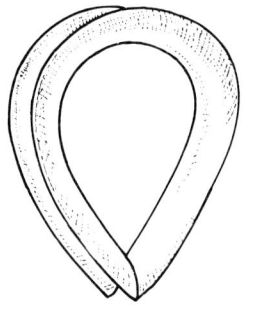

Heart – shaped thimble

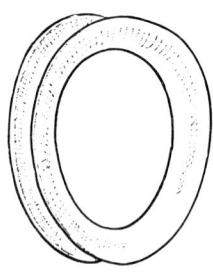

Egg – shaped thimble

Bulldog grip

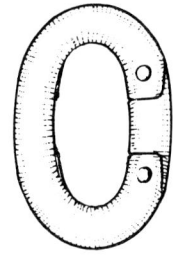

False link

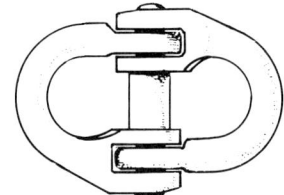

Draglink connector

Fig 25   Eyes, links and grips

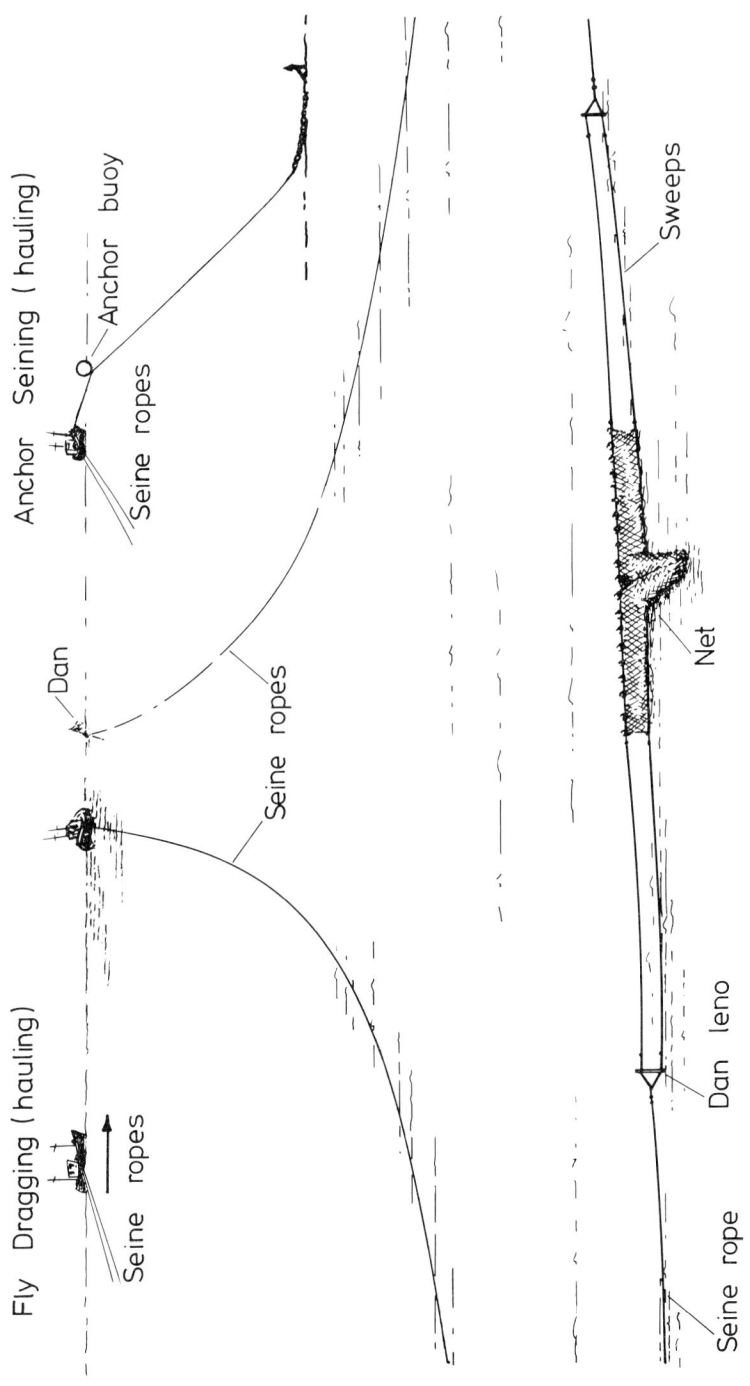

Fly Dragging (hauling)

Seine ropes

Dan

Seine ropes

Anchor Seining (hauling)

Anchor buoy

Seine ropes

Seine ropes

Sweeps

Net

Dan leno

Seine rope

Fig 26 Principal features of a Danish seine for fly dragging and anchor seining

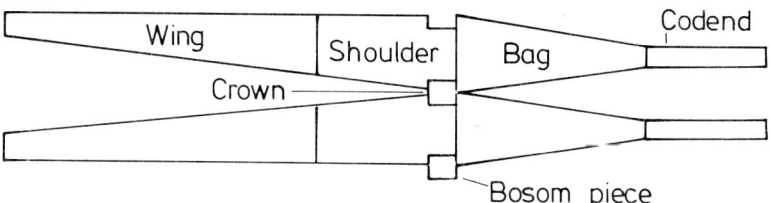

The prinipal net sections of a traditional seine net

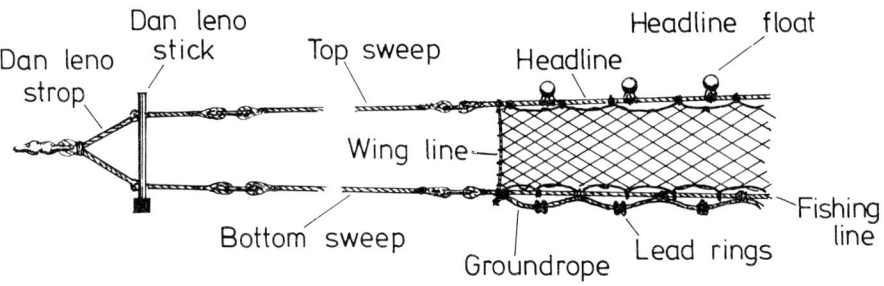

Dan leno hoop with swivel   Dan leno triangle

Fig 27 Danish seine

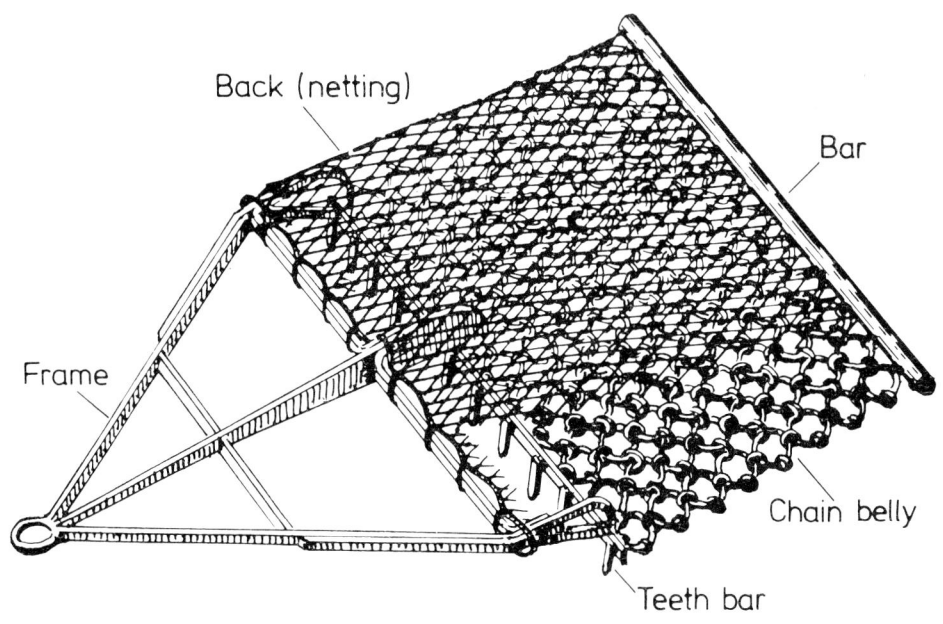

Back (netting)

Bar

Frame

Chain belly

Teeth bar

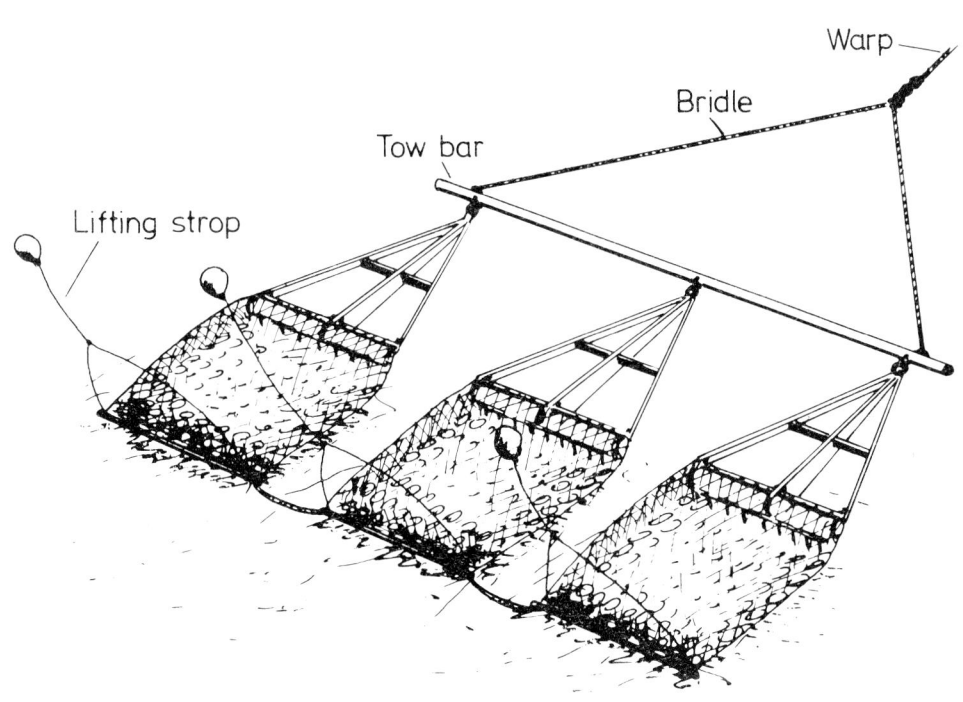

Warp

Bridle

Tow bar

Lifting strop

Fig 28 Dredges (for scallops and queens)

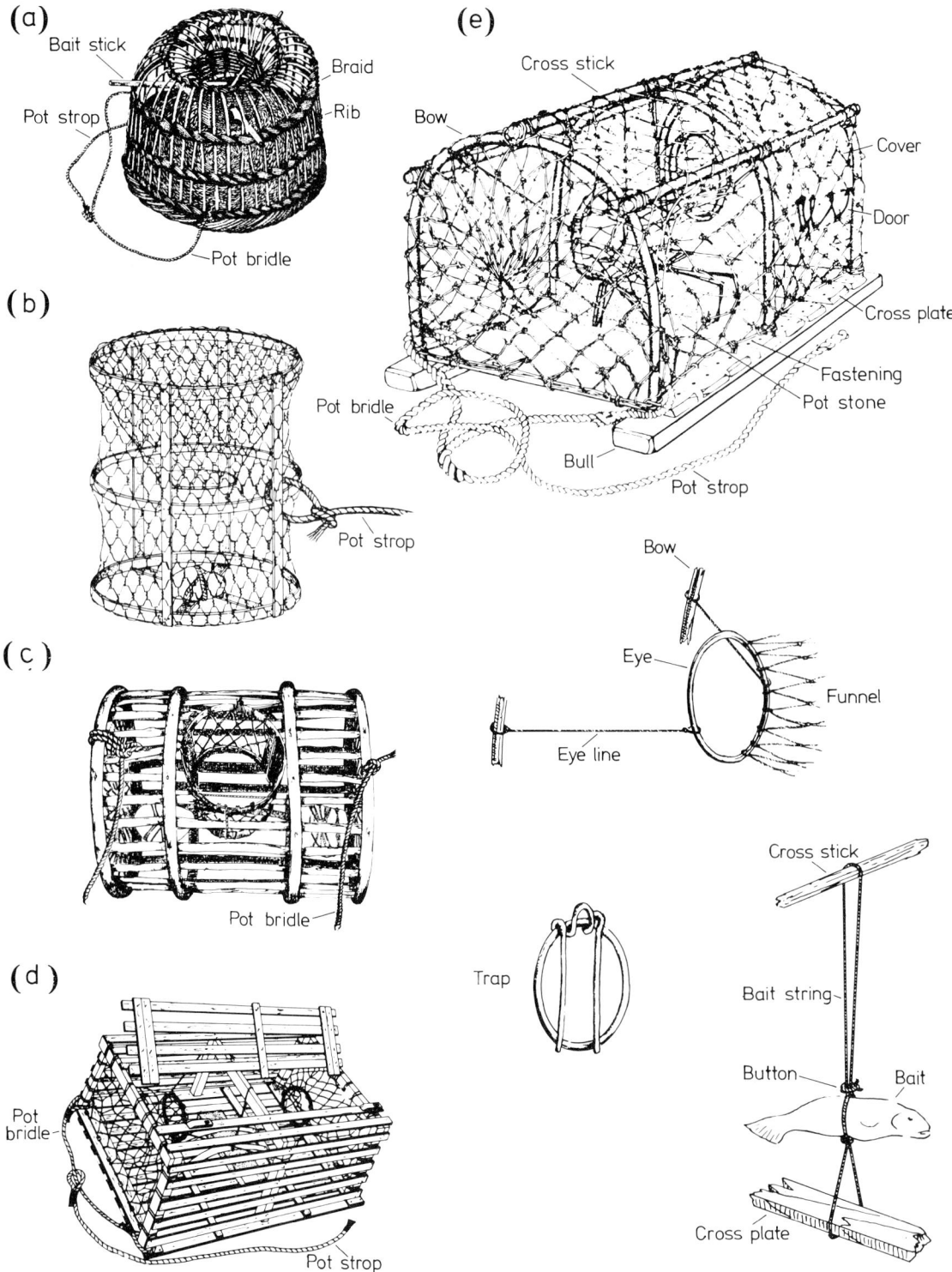

Fig 29 Shellfish pots—(a) Cornish pot, (b) Dutch pot, (c) French pot, (d) Gourdon creel, (e) Scottish creel

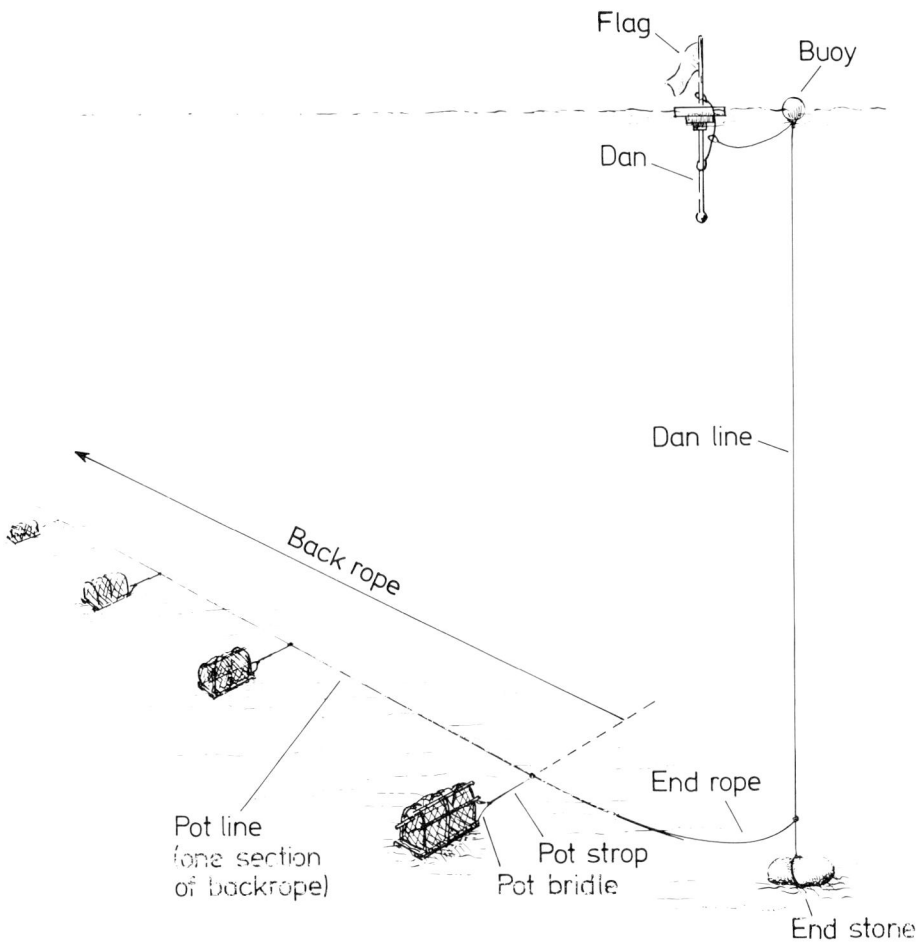

Flag

Buoy

Dan

Dan line

Back rope

Pot line
(one section
of backrope)

Pot strop

Pot bridle

End rope

End stone

Fig 30 String of pots

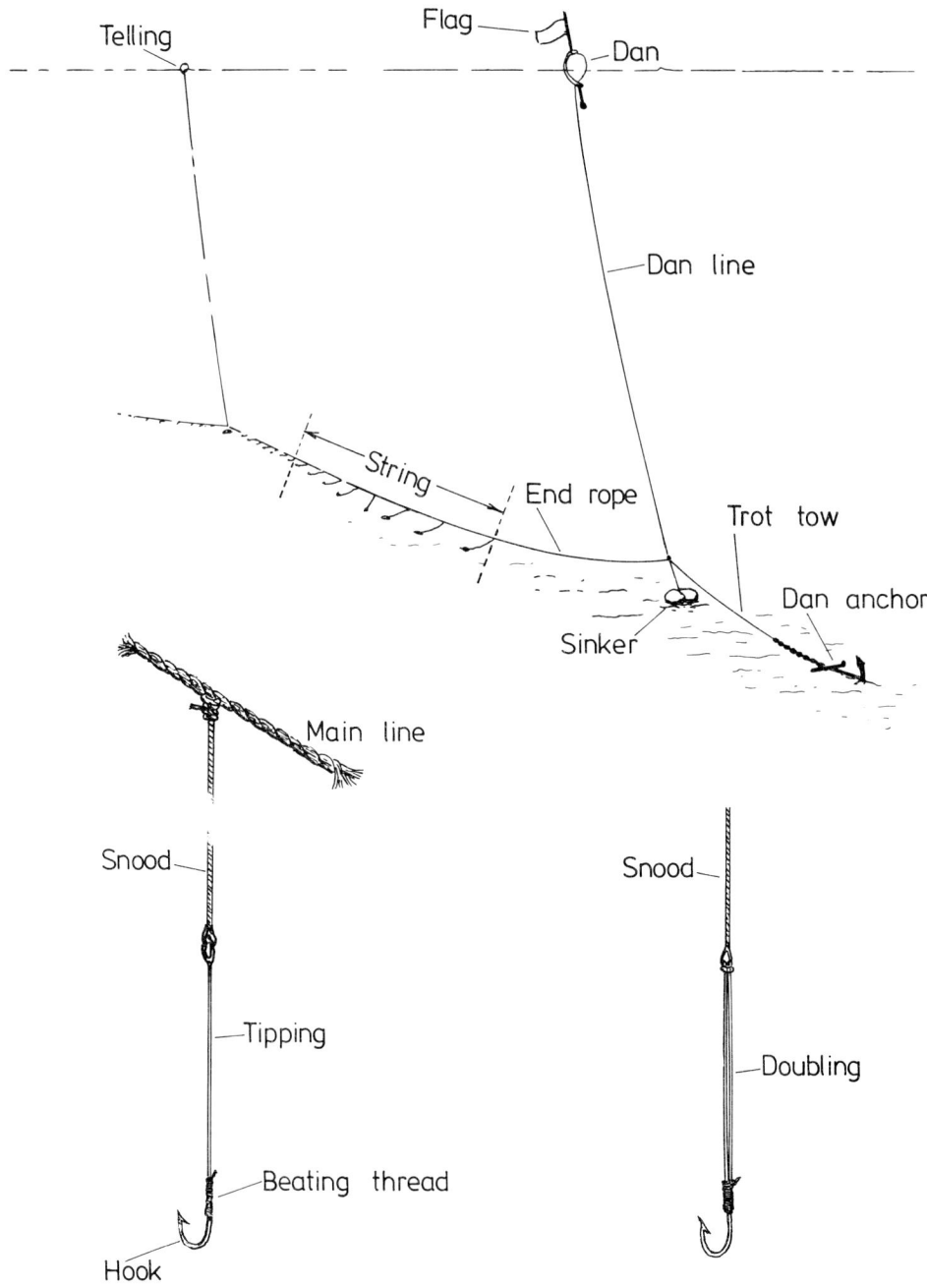

Telling

Flag — Dan

Dan line

String

End rope

Trot tow

Sinker

Dan anchor

Main line

Snood

Snood

Tipping

Doubling

Beating thread

Hook

Fig 31 Longline

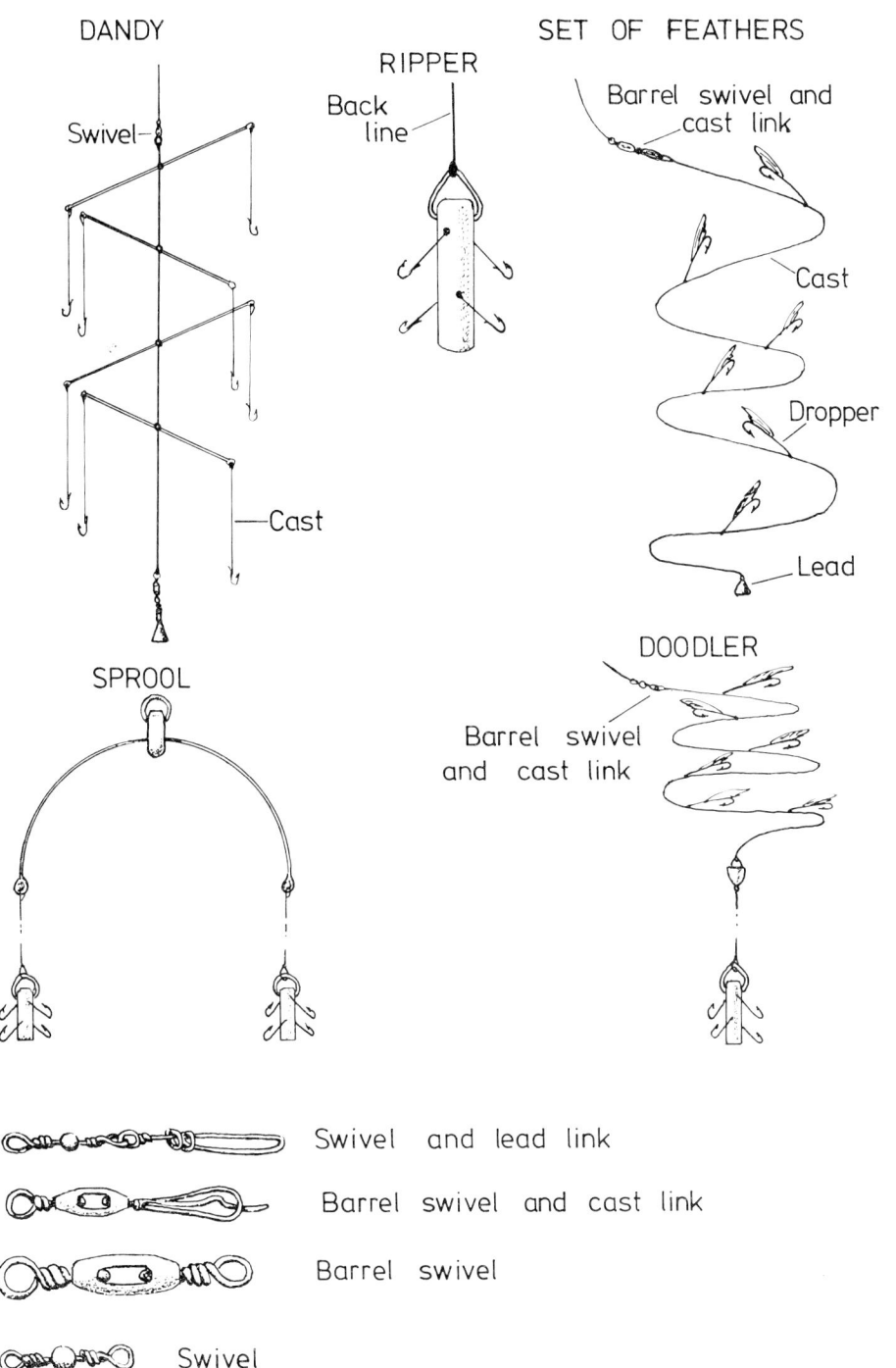

DANDY

Swivel

Cast

RIPPER

Back
line

SET OF FEATHERS

Barrel swivel and
cast link

Cast

Dropper

Lead

SPROOL

DOODLER

Barrel swivel
and cast link

Swivel and lead link

Barrel swivel and cast link

Barrel swivel

Swivel

Fig 32 Handline tackle

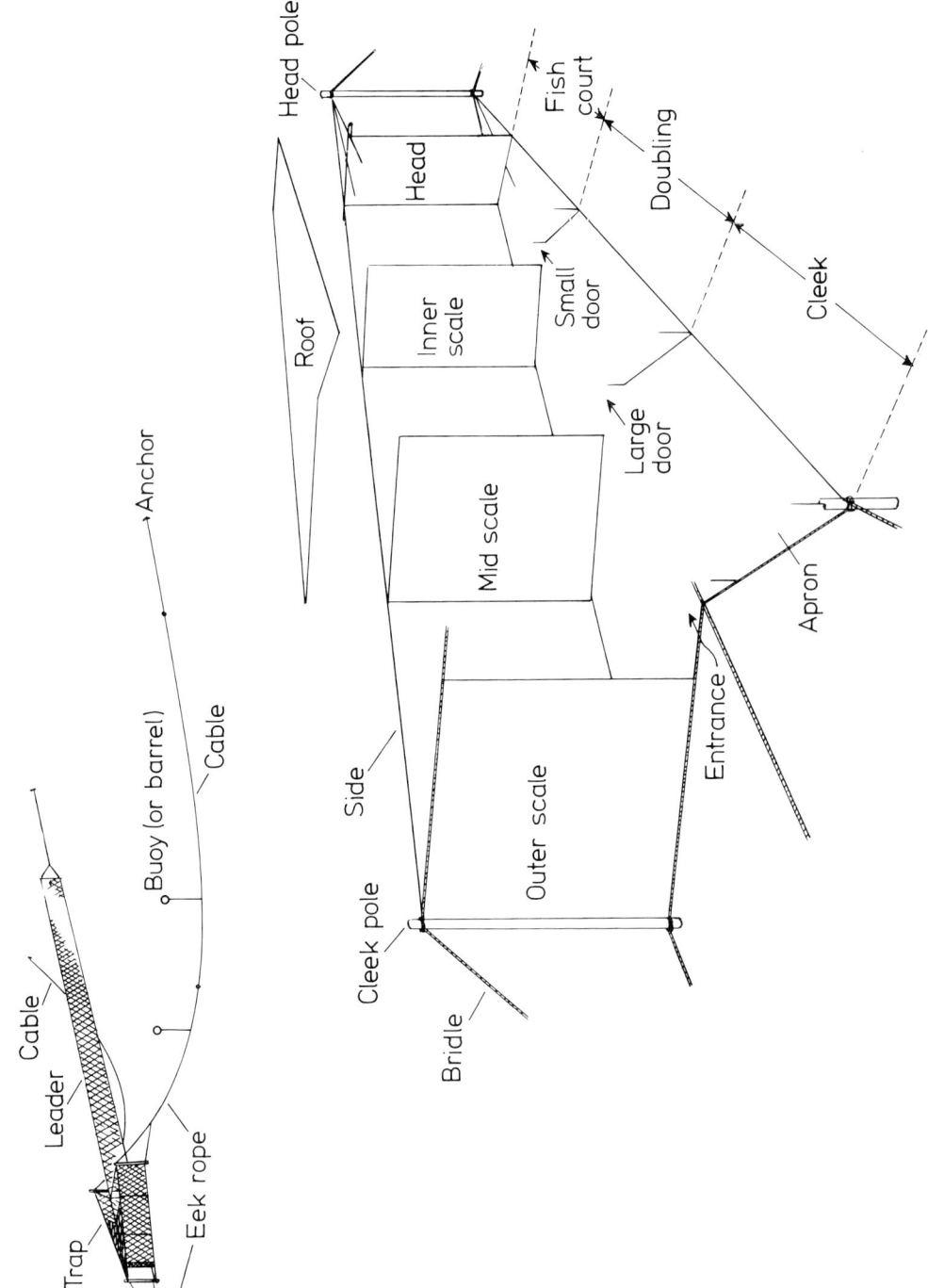

Anchor

Buoy (or barrel)

Cable

Cable

Leader

Trap

Bridle

Eek rope

Head pole

Fish court

Doubling

Head

Cleek

Roof

Inner scale

Small door

Large door

Mid scale

Side

Apron

Entrance

Outer scale

Cleek pole

Bridle

Fig 33 Bag net

Other books published by
Fishing News Books Ltd

Free catalogue available on request

Advances in aquaculture
Advances in fish science and technology
Aquaculture practices in Taiwan
Atlantic salmon: its future
Better angling with simple science
British freshwater fishes
Commercial fishing methods
Control of fish quality
Culture of bivalve molluscs
Echo sounding and sonar for fishing
The edible crab and its fishery in British waters
Eel capture, culture, processing and marketing
Eel culture
European inland water fish: a multilingual
    catalogue
FAO catalogue of fishing gear designs
FAO catalogue of small scale fishing gear
FAO investigates ferro-cement fishing craft
Farming the edge of the sea
Fish and shellfish farming in coastal waters
Fish catching methods of the world
Fisheries of Australia
Fisheries oceanography and ecology
Fishermen's handbook
Fishery products
Fishing boats and their equipment
Fishing boats of the world 1
Fishing boats of the world 2
Fishing boats of the world 3
The fishing cadet's handbook
Fishing ports and markets
Fishing with electricity
Fishing with light
Freezing and irradiation of fish
Handbook of trout and salmon diseases
Handy medical guide for seafarers
How to make and set nets
Inshore fishing: its skills, risks, rewards
Introduction to fishery by-products

The lemon sole
A living from lobsters
Marine fisheries ecosystem: its quantitative
    evaluation and management
Marine pollution and sea life
The marketing of shellfish
Mending of fishing nets
Modern deep sea trawling gear
Modern fishing gear of the world 1
Modern fishing gear of the world 2
Modern fishing gear of the world 3
More Scottish fishing craft and their work
Multilingual dictionary of fish and fish products
Naviation primer for fishermen
Netting materials for fishing gear
Pair trawling and pair seining: the technology of
    two boat fishing
Pelagic and semi-pelagic trawling gear
Planning of aquaculture development: an
    introductory guide
Power transmission and automation for ships
    and submersibles
Refrigeration on fishing vessels
Salmon and trout farming in Norway
Salmon fisheries of Scotland
Scallops and the diver-fisherman
Seafood fishing for amateur and professional
Seine fishing: bottom fishing with rope warps
    and wing trawls
Stability and trim of fishing vessels
Study of the sea
The stern trawler
Textbook of fish culture: breeding and cultivation
    of fish
Training fishermen at sea
Trout farming manual
Tuna: distribution and migration
Tuna fishing with pole and line